Houghton
Mifflin
Harcourt

Made in the United States
Text printed on 100%
recycled paper

Houghton
Mifflin
Harcourt

ISBN 978-0-544-34250-7

13 14 15 16 17   0928   22 21 20 19 18
4500713556   B C D E F G

Dear Students and Families,

Welcome to **Go Math!**, Grade 6! In this exciting mathematics program, there are hands-on activities to do and real-world problems to solve. Best of all, you will write your ideas and answers right in your book. In **Go Math!**, writing and drawing on the pages helps you think deeply about what you are learning, and you will really understand math!

By the way, all of the pages in your **Go Math!** book are made using recycled paper. We wanted you to know that you can Go Green with **Go Math!**

Sincerely,

The Authors

Made in the United States
Text printed on 100% recycled paper

GO MATH!

Authors

Juli K. Dixon, Ph.D.
Professor, Mathematics Education
University of Central Florida
Orlando, Florida

Edward B. Burger, Ph.D.
President, Southwestern University
Georgetown, Texas

Steven J. Leinwand
Principal Research Analyst
American Institutes for
    Research (AIR)
Washington, D.C.

Contributor

Rena Petrello
Professor, Mathematics
Moorpark College
Moorpark, California

Matthew R. Larson, Ph.D.
K-12 Curriculum Specialist for
    Mathematics
Lincoln Public Schools
Lincoln, Nebraska

Martha E. Sandoval-Martinez
Math Instructor
El Camino College
Torrance, California

English Language Learners Consultant

Elizabeth Jiménez
CEO, GEMAS Consulting
Professional Expert on English
    Learner Education
Bilingual Education and
    Dual Language
Pomona, California

Geometry and Statistics

Common Core **Critical Area** Solve real-world and mathematical problems involving area, surface area, and volume; and developing understanding of statistical thinking

10 Area   **531**

COMMON CORE STATE STANDARDS

6.G Geometry
Cluster A Solve real-world and mathematical problems involving area, surface area, and volume.
6.G.A.1, 6.G.A.3

GO DIGITAL

Go online! Your math lessons are interactive. Use *i*Tools, Animated Math Models, the Multimedia eGlossary, and more.

Chapter 10 Overview

In this chapter, you will explore and discover answers to the following **Essential Questions**:

- How can you use measurements to describe two-dimensional figures?
- What does area represent?
- How are the areas of rectangles and parallelograms related?
- How are the areas of triangles and trapezoids related?

Personal Math Trainer
Online Assessment and Intervention

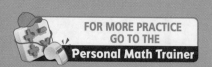

FOR MORE PRACTICE
GO TO THE
Personal Math Trainer

Practice and Homework

Lesson Check and
Spiral Review in
every lesson

Critical Area Geometry and Statistics

CRITICAL AREA
Solve real-world and mathematical problems involving area, surface area, and volume.

Developing understanding of statistical thinking

The San Francisco zoo in San Francisco, California, is home to hundreds of different animals, including this Bengal tiger.

529

This Place is a Zoo!

Planning a zoo is a difficult task. Each animal requires a special environment with different amounts of space and different features.

Get Started

You are helping to design a new section of a zoo. The table lists some of the new attractions planned for the zoo. Each attraction includes notes about the type and the amount of space needed. The zoo owns a rectangle of land that is 100 feet long and 60 feet wide. Find the dimensions of each of the attractions and draw a sketch of the plan for the zoo.

Important Facts

Attraction	Minimum Floor Space (sq ft)	Notes
American Alligators	400	rectangular pen with one side at least 24 feet long
Amur Tigers	750	trapezoid-shaped area with one side at least 40 feet long
Howler Monkeys	450	parallelogram-shaped cage with one side at least 30 feet long
Meerkat Village	250	square pen with glass sides
Red Foxes	350	rectangular pen with length twice as long as width
Tropical Aquarium	200	triangular bottom with base at least 20 feet long

_____

_____

_____

_____

_____

Completed by _____

 Show What You Know

Personal Math Trainer
Online Assessment
and Intervention

Check your understanding of important skills.

Name _____

▶ **Perimeter** **Find the perimeter.** (4.MD.A.3)

1.

$P =$ _____ units

2.

8 mm   15 mm

17 mm

$P =$ _____ mm

▶ **Identify Polygons** **Name each polygon based on the number of sides.** (5.G.B.4)

3.

_____

4.

_____

5.

_____

▶ **Evaluate Algebraic Expressions** **Evaluate the expression.** (6.EE.A.2c)

6. $5x + 2y$ for $x = 7$
and $y = 9$

_____

7. $6a \times 3b + 4$ for $a = 2$
and $b = 8$

_____

8. $s^2 + t^2 - 2^3$ for $s = 4$
and $t = 6$

_____

 **Math in the Real World**

Ross needs to paint the white boundary lines of one end zone on a football field. The area of the end zone is 4,800 square feet, and one side of the end zone measures 30 feet. One can of paint is enough to paint 300 feet of line. Help Ross find out if one can is enough to line the perimeter of the end zone.

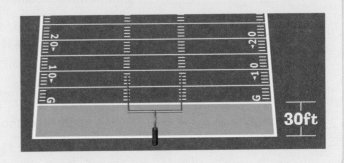

30ft

Vocabulary Builder

▶ **Visualize It** •••••••••••••••••••••••••••••••

Complete the bubble map by using the checked words that are types of quadrilaterals.

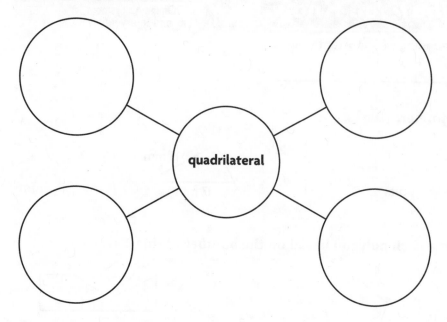

quadrilateral

▶ **Understand Vocabulary** ••••••••••••••••••••••

Complete the sentences using the preview words.

1. The _____ of a figure is the measure of the number of unit squares needed to cover it without any gaps or overlaps.

2. A polygon in which all sides are the same length and all angles have the same measure is called a(n) _____.

3. A(n) _____ is a quadrilateral with at least one pair of parallel sides.

4. _____ figures have the same size and shape.

5. A quadrilateral with two pairs of parallel sides is called a

   _____.

6. A(n) _____ is made up of more than one shape.

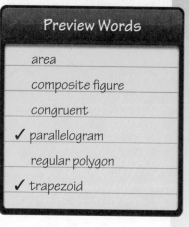

• **Interactive Student Edition**
• **Multimedia eGlossary**

Chapter 10 Vocabulary

area

área

4

composite figure

figura compuesta

14

congruent

congruente

15

parallelogram

paralelogramo

73

polygon

polígono

75

quadrilateral

cuadrilátero

83

regular polygon

polígono regular

90

trapezoid

trapecio

102

A figure that is made up of two or more simpler figures, such as triangles and quadrilaterals

Example:

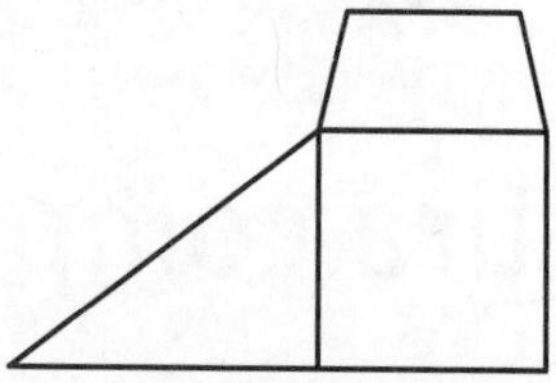

The measure of the number of unit squares needed to cover a surface without any gaps or overlaps

Example:

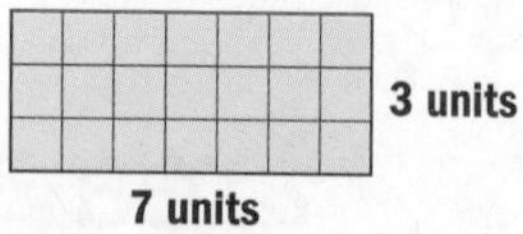

3 units

7 units

Area = 21 square units

A quadrilateral whose opposite sides are parallel and congruent

Example:

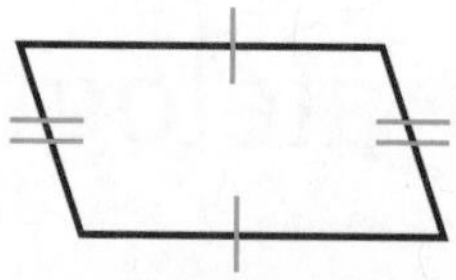

Having the same size and shape

Example:

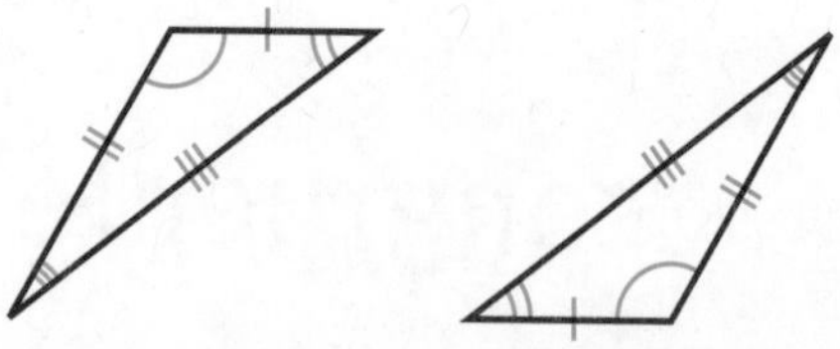

A polygon with four sides and four angles

Example:

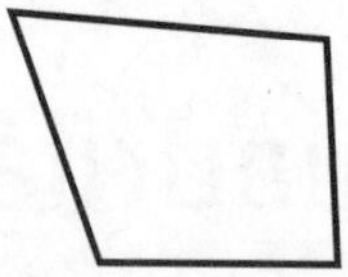

A closed plane figure formed by three or more line segments

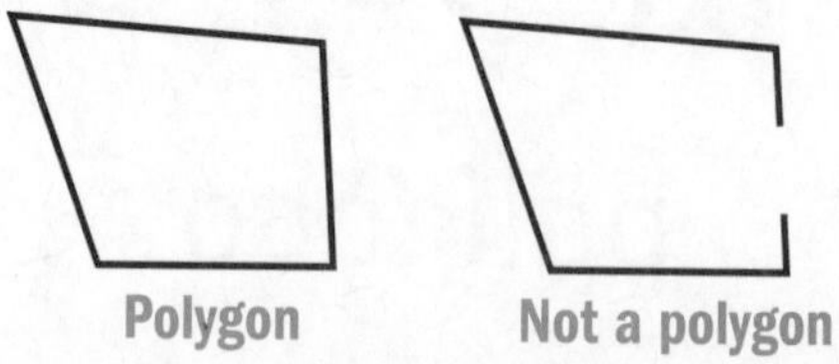

Polygon   Not a polygon

A quadrilateral with at least one pair of parallel sides

Examples:

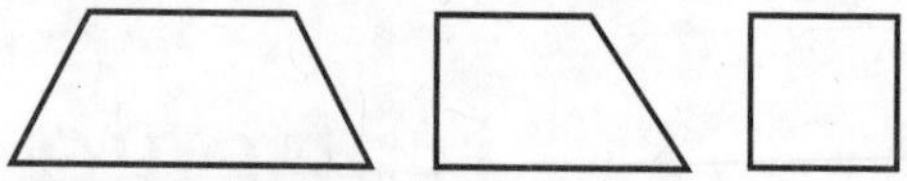

A polygon in which all sides are congruent and all angles are congruent

Example:

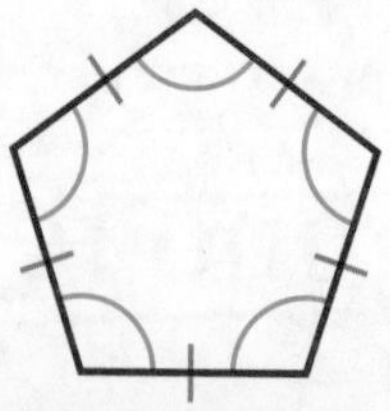

Going to the Philadelphia Zoo

Word Box
- area
- composite figure
- congruent
- parallelogram
- polygon
- quadrilateral
- regular polygon
- trapezoid

For 2 players

Materials
- 1 each: playing pieces
- 1 number cube

How to Play

1. Each player chooses a playing piece and puts it on START.
2. Toss the number cube to take a turn. Move your playing piece that many spaces.
3. If you land on these spaces:

   Light Green Tell the meaning of the math term or use it in a sentence. If the other player agrees that your answer is correct, jump to the next space with the same term.

   Dark Green Follow the directions printed in the space. If there are no directions, stay where you are.

4. The first player to reach FINISH wins.

DIRECTIONS Each player chooses a playing piece and puts it on START. • Toss the number cube to take a turn. Move your playing piece that many spaces. • If you land on these spaces: Light Green Tell the meaning of the math term or use it in a sentence. If the other player agrees that your answer is correct, jump to the next space with the same term. • Dark Green Follow the directions printed in the space. If there are no directions, stay where you are. • The first player to reach FINISH wins.

FINISH

polygon | parallelogram | congruent | composite figure

regular polygon | trapezoid | Go back to | area | composite figure

addends | quadrilateral | polygon | parallelogram

regular polygon | trapezoid | area | composite figure | congruent

addends | quadrilateral | polygon | parallelogram

START

area | composite figure | congruent | parallelogram

area

Go back to

trapezoid

regular polygon

Go back to

congruent

parallelogram

polygon

quadrilateral

addends

congruent

composite figure

area

trapezoid

regular polygon

Go back to

parallelogram

polygon

quadrilateral

addends

congruent

composite figure

area

trapezoid

polygon

quadrilateral

addends

regular polygon

Image Credits: (bg) ©Digital Vision/Getty Images; (elephant) ©Elvele Images Ltd/Alamy; (lion) ©PhotoDisc/Getty Images; (snake) ©Design Pics/Superstock

 **Journal**

The Write Way

Reflect

Choose one idea. Write about it.

- Explain how to find the area of a parallelogram.
- Tell how a trapezoid and a quadrilateral are related.
- Write a paragraph that uses the following words

  area   congruent   regular polygon
- Describe three real-life objects that are composite figures.

Area of Parallelograms

Essential Question How can you find the area of parallelograms?

Common Core Geometry—**6.G.A.1** *Also*
6.EE.A.2c, 6.EE.B.7
MATHEMATICAL PRACTICES
MP4, MP5, MP6

CONNECT The **area** of a figure is the measure of the number of unit squares needed to cover it without any gaps or overlaps. The area of a rectangle is the product of the length and the width. The rectangle shown has an area of 12 square units. For a rectangle with length l and width w, $A = l \times w$, or $A = lw$.

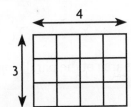

Recall that a rectangle is a special type of parallelogram. A parallelogram is a quadrilateral with two pairs of parallel sides.

Unlock the Problem

Victoria is making a quilt. She is using material in the shape of parallelograms to form the pattern. The base of each parallelogram measures 9 cm and the height measures 4 cm. What is the area of each parallelogram?

Activity Use the area of a rectangle to find the area of the parallelogram.

Materials ■ grid paper ■ scissors

- Draw the parallelogram on grid paper and cut it out.

- Cut along the dashed line to remove a triangle.

- Move the triangle to the right side of the figure to form a rectangle.

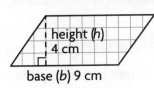

- What is the area of the rectangle? _____

- What is the area of the parallelogram? _____

- base of parallelogram = _____ of rectangle

  height of parallelogram = _____ of rectangle

  area of parallelogram = _____ of rectangle

- For a parallelogram with base b and height h, $A =$ _____

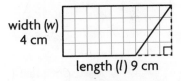

Area of parallelogram = $b \times h$ = 9 cm × 4 cm = _____ sq cm

So, the area of each parallelogram in the quilt is _____ sq cm.

> **Math Idea**
> The height of a parallelogram forms a 90° angle with the base.

Math Talk

MATHEMATICAL PRACTICES ⑥

Explain how you know that the area of the parallelogram is the same as the area of the rectangle.

🔑 Example 1 Use the formula $A = bh$ to find the area of the parallelogram.

Write the formula. $A = bh$

Replace b and h with their values. $A = 6.3 \times$ _____

Multiply. $A =$ _____

So, the area of the parallelogram is _____ square meters.

2.1 m

6.3 m

A square is a special rectangle in which the length and width are equal. For a square with side length s, $A = l \times w = s \times s = s^2$, or $A = s^2$.

🔑 Example 2 Find the area of a square with sides measuring 9.5 cm.

Write the formula. $A = s^2$

Substitute 9.5 for s. Simplify. $A = ($_____$)^2 =$ _____

So, the area of the square is _____ cm^2.

9.5 cm

9.5 cm

🔑 Example 3 A parallelogram has an area of 98 square feet and a base of 14 feet. What is the height of the parallelogram?

Write the formula. $A = bh$

Replace A and b with their values.   _____ = _____ $\times h$

Use the Division Property of Equality. $\dfrac{98}{} = \dfrac{14h}{}$

Solve for h.   _____ $= h$

So, the height of the parallelogram is _____ feet.

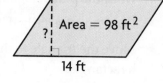

Area = 98 ft^2

14 ft

- **MATHEMATICAL PRACTICE** ⑥ **Compare** Explain the difference between the height of a rectangle and the height of a parallelogram.

_____

_____

_____

Name _____

Find the area of the parallelogram or square.

1. $A = bh$

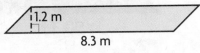

$A = 8.3 \times 1.2$

$A =$ _____ m²

2.

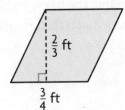

6 ft

15 ft

_____ ft²

3. 2.5 mm

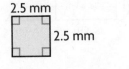

2.5 mm

_____ mm²

4.

$\frac{2}{3}$ ft

$\frac{3}{4}$ ft

_____ ft²

Find the unknown measurement for the parallelogram.

5. Area = 11 yd²

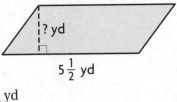

? yd

$5\frac{1}{2}$ yd

_____ yd

6. Area = 32 yd²

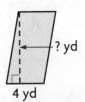

? yd

4 yd

_____ yd

Math Talk MATHEMATICAL PRACTICES ②

Reasoning Explain how the areas of some parallelograms and rectangles are related.

On Your Own

Find the area of the parallelogram.

7.

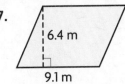

6.4 m

9.1 m

_____ m²

8.

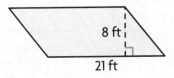

8 ft

21 ft

_____ ft²

Find the unknown measurement for the figure.

9. square

$A =$ _____

$s = 15$ ft

10. parallelogram

$A = 32$ m²

$b =$ _____

$h = 8$ m

11. parallelogram

$A = 51\frac{1}{4}$ in.²

$b = 8\frac{1}{5}$ in.

$h =$ _____

12. parallelogram

$A = 121$ mm²

$b = 11$ mm

$h =$ _____

13. THINK SMARTER The height of a parallelogram is four times the base. The base measures $3\frac{1}{2}$ ft. Find the area of the parallelogram.

_____

Problem Solving • Applications

14. Jane's backyard is shaped like a parallelogram. The base of the parallelogram is 90 feet, and the height is 25 feet. What is the area of Jane's backyard?

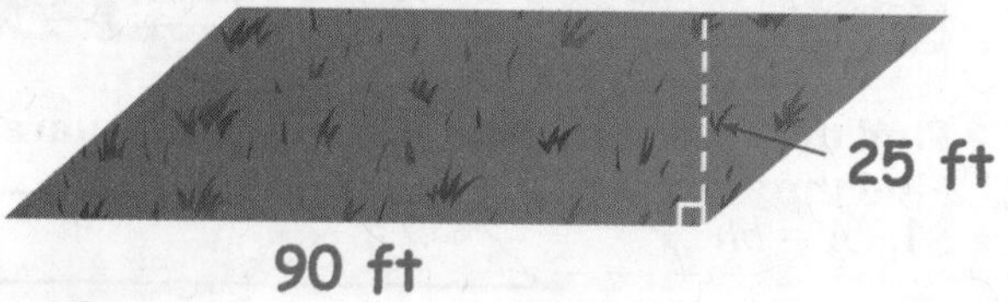

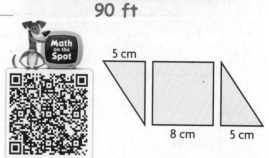

_____

15. **THINK SMARTER** Jack made a parallelogram by putting together two congruent triangles and a square, like the figures shown at the right. The triangles have the same height as the square. What is the area of Jack's parallelogram?

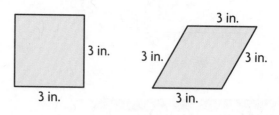

_____

16. **GO DEEPER** The base of a parallelogram is 2 times the parallelogram's height. If the base is 12 inches, what is the area?

_____

17. **MATHEMATICAL PRACTICE ③** **Verify the Reasoning of Others** Li Ping says that a square with 3-inch sides has a greater area than a parallelogram that is not a square but has sides that have the same length. Does Li Ping's statement make sense? Explain.

_____

_____

_____

_____

18. **THINK SMARTER** Find the area of the parallelogram.

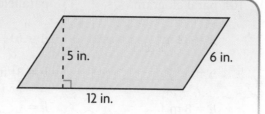

The area is _____ in².

Name _____

Area of Parallelograms

Common Core **COMMON CORE STANDARD—6.G.A.1**
Solve real-world and mathematical problems involving area, surface area, and volume.

Find the area of the figure.

1.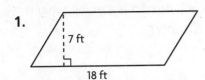
7 ft
18 ft

$A = bh$
$A = 18 \times 7$
$A = \textbf{126 ft}^2$

2.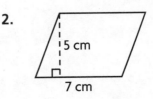
5 cm
7 cm

_____ cm²

Find the unknown measurement for the figure.

3. parallelogram

$A = 9.18 \text{ m}^2$

$b = 2.7 \text{ m}$

$h = \underline{\hspace{1cm}}$

4. parallelogram

$A = \underline{\hspace{1cm}}$

$b = 4\frac{3}{10}\text{m}$

$h = 2\frac{1}{10}\text{m}$

5. square

$A = \underline{\hspace{1cm}}$

$s = 35 \text{ cm}$

6. parallelogram

$A = 6.3 \text{ mm}^2$

$b = \underline{\hspace{1cm}}$

$h = 0.9 \text{ mm}$

Problem Solving · Real World

7. Ronna has a sticker in the shape of a parallelogram. The sticker has a base of 6.5 cm and a height of 10.1 cm. What is the area of the sticker?

8. A parallelogram-shaped tile has an area of 48 in.². The base of the tile measures 12 in. What is the measure of its height?

_____

_____

9. **WRITE** ▸Math Copy the two triangles and the square in Exercise 15 on page 536. Show how you found the area of each piece. Draw the parallelogram formed when the three figures are put together. Calculate its area using the formula for the area of a parallelogram.

_____

_____

Lesson Check (6.G.A.1, 6.EE.A.2c, 6.EE.B.7)

1. Cougar Park is shaped like a parallelogram and has an area of $\frac{1}{16}$ square mile. Its base is $\frac{3}{8}$ mile. What is its height?

2. Square County is a square-shaped county divided into 16 equal-sized square districts. If the side length of each district is 4 miles, what is the area of Square County?

Spiral Review (6.EE.B.5, 6.EE.B.8, 6.EE.C.9)

3. Which of the following values of y make the inequality $y < {}^-4$ true?

   $y = {}^-4$ $y = {}^-6$ $y = 0$ $y = {}^-8$ $y = 2$

4. On a winter's day, 9°F is the highest temperature recorded. Write an inequality that represents the temperature t in degrees Fahrenheit at any time on this day.

5. In 2 seconds, an elevator travels 40 feet. In 3 seconds, the elevator travels 60 feet. In 4 seconds, the elevator travels 80 feet. Write an equation that gives the relationship between the number of seconds x and the distance y the elevator travels.

6. The linear equation $y = 4x$ represents the number of bracelets y that Jolene can make in x hours. Which ordered pair lies on the graph of the equation?

FOR MORE PRACTICE
GO TO THE
Personal Math Trainer

Name _____

Explore Area of Triangles

Essential Question What is the relationship among the areas of triangles, rectangles, and parallelograms?

Common Core Geometry—
6.G.A.1
MATHEMATICAL PRACTICES
MP1, MP3, MP5, MP8

Investigate

Materials ■ grid paper ■ tracing paper ■ ruler ■ scissors

A. On the grid, draw a rectangle with a base of 6 units and a height of 5 units.

- What is the area of the rectangle?

_____

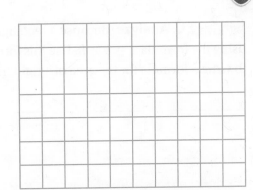

B. Trace the rectangle onto tracing paper. Draw a diagonal from the top-left corner to the lower-right corner.

- A diagonal is a line segment that connects two nonadjacent vertices of a polygon.

C. Cut out the rectangle. Then cut along the diagonal to divide the rectangle into two right triangles. Compare the two triangles.

- **Congruent** figures are the same shape and size. Are the two right triangles congruent?

_____

_____

- How is the area of each right triangle related to the area of the rectangle?

_____

_____

- What is the area of each right triangle?

_____

Draw Conclusions

1. Explain how finding the area of a rectangle is like finding the area of a right triangle. How is it different?

_____

_____

2. 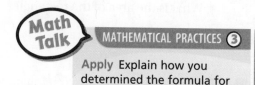 **Analyze** Because a rectangle is a parallelogram, its area can be found using the formula $A = b \times h$. Use this formula and your results from the Investigate to write a formula for the area of a right triangle with base b and height h.

_____

Math Talk

MATHEMATICAL PRACTICES ③

Apply Explain how you determined the formula for the area of a right triangle.

Make Connections

The area of any parallelogram, including a rectangle, can be found using the formula $A = b \times h$. You can use a parallelogram to look at more triangles.

A. Trace and cut out two copies of the acute triangle.

B. Arrange the two triangles to make a parallelogram.

- Are the triangles congruent? _____

- If the area of the parallelogram is 10 square centimeters, what is the area of each triangle? Explain how you know.

_____

_____

_____

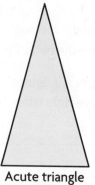

Acute triangle

C. Repeat Steps A and B with the obtuse triangle.

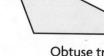

Obtuse triangle

3. MATHEMATICAL PRACTICE ⑧ **Generalize** Can you use the formula $A = \frac{1}{2} \times b \times h$ to find the area of any triangle? Explain.

_____

_____

Name _____

Share and Show MATH BOARD

1. Trace the parallelogram, and cut it into two congruent triangles. Find the areas of the parallelogram and one triangle, using square units.

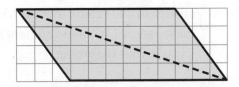

_____

Find the area of each triangle.

2.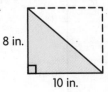
8 in.
10 in.

_____ in.²

✓ 3.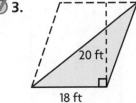
20 ft
18 ft

_____ ft²

✓ 4.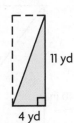
11 yd
4 yd

_____ yd²

5.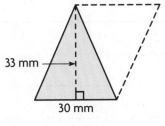
33 mm →
30 mm

_____ mm²

6.
20 in.
19 in.

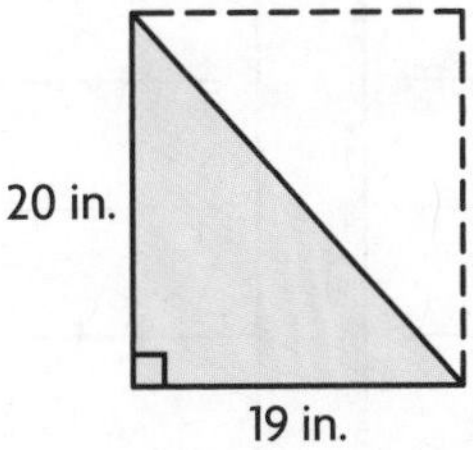

_____ in.²

7.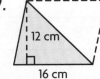
12 cm
16 cm

_____ cm²

Problem Solving • Applications Real World

8. **MATHEMATICAL PRACTICE ⑤** **Communicate** Describe how you can use two triangles of the same shape and size to form a parallelogram.

_____

_____

_____

9. **GO DEEPER** A school flag is in the shape of a right triangle. The height of the flag is 36 inches and the base is $\frac{3}{4}$ of the height. What is the area of the flag?

_____

THINK SMARTER **Sense or Nonsense?**

10. Cyndi and Tyson drew the models below. Each said his or her drawing represents a triangle with an area of 600 square inches. Whose statement makes sense? Whose statement is nonsense? Explain your reasoning.

Tyson's Model

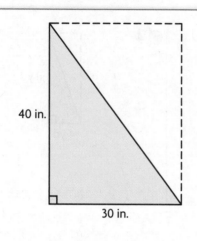

40 in.

30 in.

_____

_____

_____

_____

Cyndi's Model

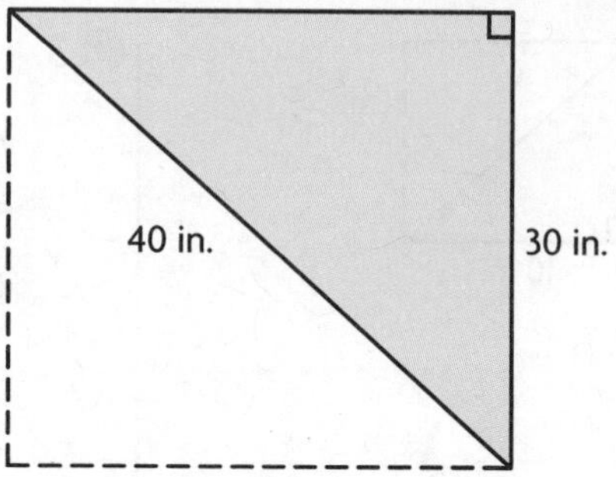

40 in.   30 in.

_____

_____

_____

_____

11. **THINK SMARTER** A flag is separated into two different colors. Find the area of the white region. Show your work.

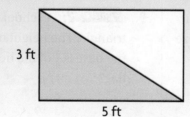

3 ft

5 ft

_____

_____

Explore Area of Triangles

 COMMON CORE STANDARD—6.G.A.1
*Solve real-world and mathematical problems
involving area, surface area, and volume.*

Find the area of each triangle.

1.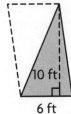
10 ft
6 ft

_____30 ft²_____

2.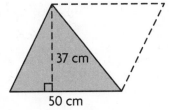
37 cm
50 cm

_____

3.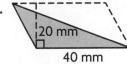
20 mm
40 mm

_____

4.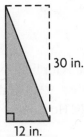
30 in.
12 in.

_____

5.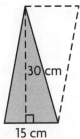
30 cm
15 cm

_____

6.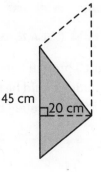
45 cm 20 cm

_____

Problem Solving · Real World

7. Fabian is decorating a triangular pennant for a football game. The pennant has a base of 10 inches and a height of 24 inches. What is the total area of the pennant?

_____

8. Ryan is buying a triangular tract of land. The triangle has a base of 100 yards and a height of 300 yards. What is the area of the tract of land?

_____

9. **WRITE** ▸*Math* Draw 3 triangles on grid paper. Draw appropriate parallelograms to support the formula for the area of the triangle. Tape your drawings to this page.

_____

Lesson Check (6.G.A.1)

1. What is the area of a triangle with a height of 14 feet and a base of 10 feet?

2. What is the area of a triangle with a height of 40 millimeters and a base of 380 millimeters?

Spiral Review (6.EE.A.2c, 6.EE.B.7, 6.EE.B.8, 6.G.A.1)

3. Jack bought 3 protein bars for a total of $4.26. Which equation could be used to find the cost c in dollars of each protein bar?

4. Coach Herrera is buying tennis balls for his team. He can solve the equation $4c = 92$ to find how many cans c of balls he needs. How many cans does he need?

5. Sketch the graph of $y \leq {}^-7$ on a number line.

6. A square photograph has a perimeter of 20 inches. What is the area of the photograph?

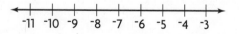

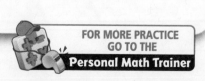

FOR MORE PRACTICE
GO TO THE
Personal Math Trainer

Name _____

Area of Triangles

Essential Question How can you find the area of triangles?

Common Core Geometry—6.G.A.1
Also 6.EE.A.2c
MATHEMATICAL PRACTICES
MP3, MP6, MP8

Any parallelogram can be divided into two congruent triangles. The area of each triangle is half the area of the parallelogram, so the area of a triangle is half the product of its base and its height.

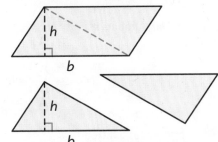

> ### Area of a Triangle
>
> $$A = \frac{1}{2} bh$$
>
> where b is the base and h is the height

Unlock the Problem

The Flatiron Building in New York is well known for its unusual shape. The building was designed to fit the triangular plot of land formed by 22nd Street, Broadway, and Fifth Avenue. The diagram shows the dimensions of the triangular foundation of the building. What is the area of the triangle?

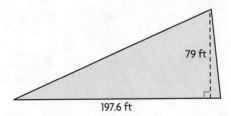

79 ft

197.6 ft

- How can you identify the base and the height of the triangle?

_____

_____

_____

_____

Find the area of the triangle.

Write the formula.

$$A = \frac{1}{2} bh$$

Substitute 197.6 for b and 79 for h.

$$A = \frac{1}{2} \times \rule{2cm}{0.4pt} \times \rule{2cm}{0.4pt}$$

Multiply the base and height.

$$A = \frac{1}{2} \times \rule{2cm}{0.4pt}$$

Multiply by $\frac{1}{2}$.

$$A = \rule{2cm}{0.4pt}$$

So, the area of the triangle is _____ ft².

Math Talk MATHEMATICAL PRACTICES ⑧

Generalize How does the area of a triangle relate to the area of a rectangle with the same base and height?

🔓 Example 1 Find the area of the triangle.

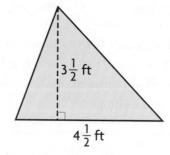

Write the formula.

$$A = \frac{1}{2}bh$$

Substitute $4\frac{1}{2}$ for b and $3\frac{1}{2}$ for h.

$$A = \frac{1}{2} \times \underline{\hspace{1.5cm}} \times \underline{\hspace{1.5cm}}$$

Rewrite the mixed numbers as fractions.

$$A = \frac{1}{2} \times \frac{\boxed{}}{2} \times \frac{\boxed{}}{2}$$

Multiply.

$$A = \frac{\boxed{}}{8}$$

Rewrite the fraction as a mixed number.

$$A = \underline{\hspace{2cm}}$$

So, the area of the triangle is _____ ft².

🔓 Example 2

Daniella is decorating a triangular pennant for her wall. The area of the pennant is 225 in.² and the base measures 30 in. What is the height of the triangular pennant?

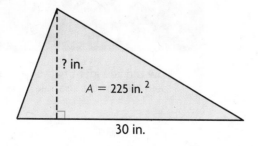

Write the formula.

$$A = \frac{1}{2}bh$$

Substitute 225 for A and 30 for b.

$$\underline{\hspace{1.5cm}} = \frac{1}{2} \times \underline{\hspace{1.5cm}} \times h$$

Multiply $\frac{1}{2}$ and 30.

$$225 = \underline{\hspace{1.5cm}} \times h$$

Use the Division Property of Equality.

$$\frac{225}{\boxed{}} = \frac{\boxed{} \times h}{\boxed{}}$$

Simplify.

$$\underline{\hspace{1.5cm}} = h$$

So, the height of the triangular pennant is _____ in.

Name ______________________

Share and Show MATH BOARD

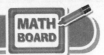

1. FInd the area of the triangle.

8 cm
14 cm

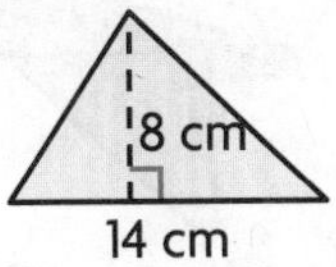

$A = \frac{1}{2}bh$

$A = \frac{1}{2} \times 14 \times$ ______

$A =$ ______ cm^2

2. The area of the triangle is 132 in.^2 Find the height of the triangle.

h
22 in.

$h =$ ______

Find the area of the triangle.

3.

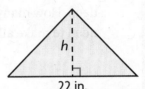

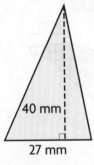

40 mm
27 mm

$A =$ ______

4.

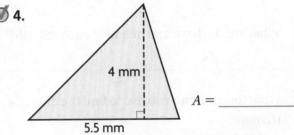

4 mm
5.5 mm

$A =$ ______

Math Talk — MATHEMATICAL PRACTICES 6

Explain how you can identify the height of a triangle.

On Your Own

THINK SMARTER **Find the unknown measurement for the figure.**

5.

Area = 52.5 in.²
5 in.
h

$h =$ ______

6.

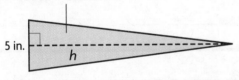

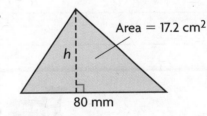

Area = 17.2 cm²
h
80 mm

$h =$ ______

7. MATHEMATICAL PRACTICE 3 **Verify the Reasoning of Others** The height of a triangle is twice the base. The area of the triangle is 625 in.^2 Carson says the base of the triangle is at least 50 in. Is Carson's estimate reasonable? Explain.

____

____

____

Unlock the Problem

8. **GO DEEPER** Alani is building a set of 4 shelves. Each shelf will have 2 supports in the shape of right isosceles triangles. Each shelf is 14 inches deep. How many square inches of wood will she need to make all of the supports?

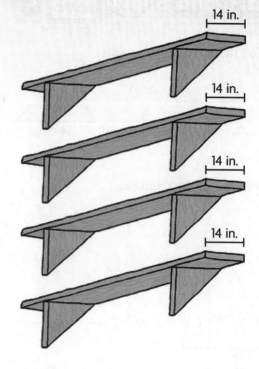

14 in.

14 in.

14 in.

14 in.

a. What are the base and height of each triangle?

_____

b. What formula can you use to find the area of a triangle?

_____

c. Explain how you can find the area of one triangular support.

_____

_____

_____

d. How many triangular supports are needed to build 4 shelves?

_____

e. How many square inches of wood will Alani need to make all the supports?

_____

---

9. **THINK SMARTER** The area of a triangle is 97.5 cm². The height of the triangle is 13 cm. Find the base of the triangle. Explain your work.

Math on the Spot

_____

_____

_____

_____

10. **THINK SMARTER** The area of a triangle is 30 ft². For numbers 10a–10d, select Yes or No to tell if the dimensions given could be the height and base of the triangle.

10a. $h = 3, b = 10$ ○ Yes   ○ No

10b. $h = 3, b = 20$ ○ Yes   ○ No

10c. $h = 5, b = 12$ ○ Yes   ○ No

10d. $h = 5, b = 24$ ○ Yes   ○ No

Name _____

Area of Triangles

Name _____

Area of Triangles

COMMON CORE STANDARD—6.G.A.1
Solve real-world and mathematical problems involving area, surface area, and volume.

Find the area.

1.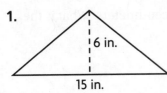
   6 in.
   15 in.

   $A = \frac{1}{2}bh$

   $A = \frac{1}{2} \times 15 \times 6$

   $A = 45$

   Area = 45 in.^{2}

2.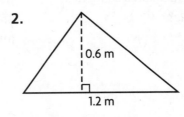
   0.6 m
   1.2 m

   _____

3.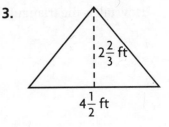
   $2\frac{2}{3}$ ft
   $4\frac{1}{2}$ ft

   _____

Find the unknown measurement for the triangle.

4. $A = 0.225$ mi^2

   $b = 0.6$ mi

   $h = $ ▨

   _____

5. $A = 4.86$ yd^2

   $b = $ ▨

   $h = 1.8$ yd

   _____

6. $A = 63$ m^2

   $b = $ ▨

   $h = 12$ m

   _____

7. $A = 2.5$ km^2

   $b = 5$ km

   $h = $ ▨

   _____

Problem Solving · Real World

8. Bayla draws a triangle with a base of 15 cm and a height of 8.5 cm. If she colors the space inside the triangle, what area does she color?

   _____

9. Alicia is making a triangular sign for the school play. The area of the sign is 558 in.^{2}. The base of the triangle is 36 in. What is the height of the triangle?

   _____

10. **WRITE** ▸*Math* Describe how you would find how much grass seed is needed to cover a triangular plot of land.

    _____

    _____

Lesson Check (6.G.A.1, 6.EE.A.2c)

1. A triangular flag has an area of 187.5 square inches. The base of the flag measures 25 inches. How tall is the triangular flag?

2. A piece of stained glass in the shape of a right triangle has sides measuring 8 centimeters, 15 centimeters, and 17 centimeters. What is the area of the piece?

Spiral Review (6.EE.B.7, 6.EE.C.9, 6.G.A.1)

3. Tina bought a t-shirt and sandals. The total cost was $41.50. The t-shirt cost $8.95. The equation $8.95 + c = 41.50$ can be used to find the cost c in dollars of the sandals. How much did the sandals cost?

4. There are 37 paper clips in a box. Carmen places more paper clips in the box. Write an equation to show the total number of paper clips p in the box after Carmen places n more paper clips in the box.

5. Name another ordered pair that is on the graph of the equation represented by the table.

People in group, x	1	2	3	4
Total cost of ordering lunch special ($), y	6	12	18	24

6. Find the area of the triangle that divides the parallelogram in half.

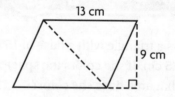

13 cm

9 cm

© Houghton Mifflin Harcourt Publishing Company

Explore Area of Trapezoids

Essential Question What is the relationship between the areas of trapezoids and parallelograms?

Common Core Geometry—
6.G.A.1

MATHEMATICAL PRACTICES
MP4, MP7, MP8

CONNECT A **trapezoid** is a quadrilateral with at least one pair of parallel sides. Any pair of parallel sides can be the *bases* of the trapezoid. A line segment drawn at a 90° angle to the two bases is the *height* of the trapezoid. You can use what you know about the area of a parallelogram to find the area of a trapezoid.

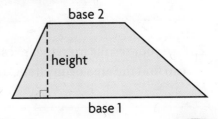

base 2

height

base 1

Investigate

Materials ■ grid paper ■ ruler ■ scissors

A. Draw two copies of the trapezoid on grid paper.

B. Cut out the trapezoids.

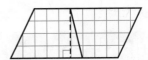

3 units

4 units

6 units

C. Arrange the trapezoids to form a parallelogram, as shown. Examine the parallelogram.

- How can you find the length of the base of the parallelogram?

  _____

  _____

- The base of the parallelogram is _____ + _____ = _____ units.

- The height of the parallelogram is _____ units.

- The area of the parallelogram is _____ × _____ = _____ square units.

D. Examine the trapezoids.

- How does the area of one trapezoid relate to the area of the parallelogram?

  _____

  _____

- Find the area of one trapezoid. Explain how you found the area.

  _____

  _____

1. **MATHEMATICAL PRACTICE ⑦** Identify Relationships Explain how knowing how to find the area of a parallelogram helped you find the area of the trapezoid.

_____

2. Use your results from the Investigate to describe how you can find the area of any trapezoid.

_____

_____

_____

3. **MATHEMATICAL PRACTICE ⑧** Generalize Can you use the method you described above to find the area of a trapezoid if two copies of the trapezoid can be arranged to form a rectangle? Explain.

_____

_____

Make Connections

You can use the formula for the area of a rectangle to find the area of some types of trapezoids.

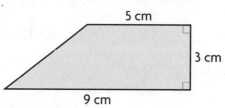

5 cm

3 cm

9 cm

A. Trace and cut out two copies of the trapezoid.

B. Arrange the two trapezoids to form a rectangle. Examine the rectangle.

- The length of the rectangle is _____ + _____ = _____ cm.

- The width of the rectangle is _____ cm.

- The area of the rectangle is _____ × _____ = _____ cm².

C. Examine the trapezoids.

- How does the area of each trapezoid relate to the area of the rectangle?

_____

- The area of the given trapezoid is $\frac{1}{2}$ × _____ = _____ cm².

Name _____

Share and Show

1. Trace and cut out two copies of the trapezoid. Arrange the trapezoids to form a parallelogram. Find the areas of the parallelogram and one trapezoid using square units.

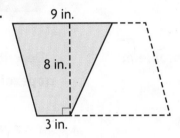

_____

Find the area of the trapezoid.

2.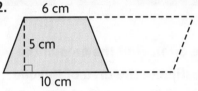
6 cm
5 cm
10 cm

_____ cm²

3.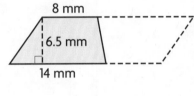
9 in.
8 in.
3 in.

_____ in.²

4.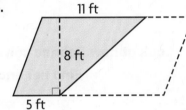
11 ft
8 ft
5 ft

_____ ft²

5.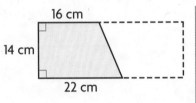
16 cm
14 cm
22 cm

_____ cm²

6.
8 mm
6.5 mm
14 mm

_____ mm²

7.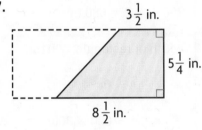
$3\frac{1}{2}$ in.
$5\frac{1}{4}$ in.
$8\frac{1}{2}$ in.

_____ in.²

Problem Solving • Applications

8. **MATHEMATICAL PRACTICE ④ Describe a Method** Explain one way to find the height of a trapezoid if you know the area of the trapezoid and the length of both bases.

_____

_____

_____

9. **GO DEEPER** A patio is in the shape of a trapezoid. The length of the longer base is 18 feet. The length of the shorter base is two feet less than half the longer base. The height is 8 feet. What is the area of the patio?

_____

_____

_____

THINK SMARTER · **What's the Error?**

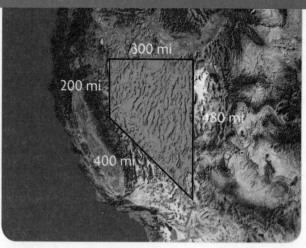

10. Except for a small region near its southeast corner, the state of Nevada is shaped like a trapezoid. The map at the right shows the approximate dimensions of the trapezoid. Sabrina used the map to estimate the area of Nevada.

Look at how Sabrina solved the problem. Find her error.

Two copies of the trapezoid can be put together to form a rectangle.

length of rectangle:

$200 + 480 = 680$ mi

width of rectangle: 300 mi

$A = lw$

$= 680 \times 300$

$= 204,000$

The area of Nevada is about 204,000 square miles.

Describe the error. Find the area of the trapezoid to estimate the area of Nevada.

---

11. THINK SMARTER · A photo was cut in half at an angle. What is the area of one of the cut pieces?

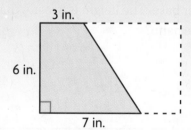

3 in.

6 in.

7 in.

The area is _____.

Name _____

Explore Area of Trapezoids

COMMON CORE STANDARD—6.G.A.1
Solve real-world and mathematical problems involving area, surface area, and volume.

1. Trace and cut out two copies of the trapezoid. Arrange the trapezoids to form a parallelogram. Find the areas of the parallelogram and the trapezoids using square units.

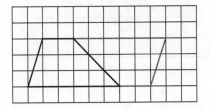

parallelogram: 24 square units; trapezoids: _____

12 square units

_____

Find the area of the trapezoid.

2.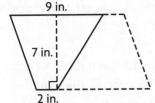

_____ in.²

3.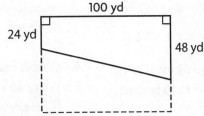

_____ yd²

4.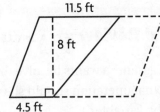

_____ ft²

Problem Solving • Real World

5. A cake is made out of two identical trapezoids. Each trapezoid has a height of 11 inches and bases of 9 inches and 14 inches. What is the area of one of the trapezoid pieces?

_____

6. A sticker is in the shape of a trapezoid. The height is 3 centimeters, and the bases are 2.5 centimeters and 5.5 centimeters. What is the area of the sticker?

_____

7. **WRITE** ▸Math Find the area of a trapezoid that has bases that are 15 inches and 20 inches and a height of 9 inches.

_____

Lesson Check (6.G.A.1)

1. What is the area of figure *ABEG*?

A ── 9 yd ── B ─────── C
| 7 yd
G F ── 15 yd ── E ──── D

2. Maggie colors a figure in the shape of a trapezoid. The trapezoid is 6 inches tall. The bases are 4.5 inches and 8 inches. What is the area of the figure that Maggie colored?

_____

_____

Spiral Review (6.EE.A.2c, 6.EE.B.7, 6.EE.C.9, 6.G.A.1)

3. Cassandra wants to solve the equation $30 = \frac{2}{5}p$. What operation should she perform to isolate the variable?

_____

_____

_____

4. Ginger makes pies and sells them for $14 each. Write an equation that represents the situation, if *y* represents the money that Ginger earns and *x* represents the number of pies sold.

5. What is the equation for the graph shown below?

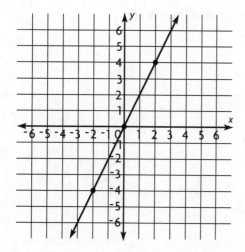

6. Cesar made a rectangular banner that is 4 feet by 3 feet. He wants to make a triangular banner that has the same area as the other banner. The triangular banner will have a base of 4 feet. What should its height be?

_____

© Houghton Mifflin Harcourt Publishing Company

FOR MORE PRACTICE
GO TO THE
Personal Math Trainer

Area of Trapezoids

Essential Question How can you find the area of trapezoids?

Any parallelogram can be divided into two trapezoids with one pair of parallel sides that are also the same shape and size. The bases of the trapezoids, b_1 and b_2, form the base of the parallelogram. The area of each trapezoid is half the area of the parallelogram. So, the area of a trapezoid is half the product of its height and the sum of its bases.

Common
Core — Geometry—6.G.A.1
Also 6.EE.A.2c
MATHEMATICAL PRACTICES
MP1, MP3, MP6

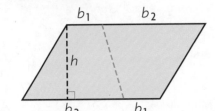

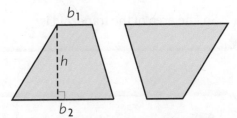

> **Area of a Trapezoid**
>
> $$A = \frac{1}{2}(b_1 + b_2)h$$
>
> where b_1 and b_2 are the two bases and h is the height

Unlock the Problem — Real World

Mr. Desmond has tables in his office with tops shaped like trapezoids. The diagram shows the dimensions of each tabletop. What is the area of each tabletop?

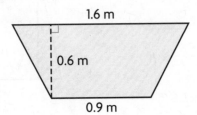

1.6 m

0.6 m

0.9 m

• How can you identify the bases?

_____

_____

• How can you identify the height?

_____

_____

Find the area of the trapezoid.

Write the formula.
$$A = \frac{1}{2}(b_1 + b_2)h$$

Substitute 1.6 for b_1, 0.9 for b_2, and 0.6 for h.
$$A = \frac{1}{2} \times (\underline{\quad} + \underline{\quad}) \times \underline{\quad}$$

Add within the parentheses.
$$A = \frac{1}{2} \times \underline{\quad} \times 0.6$$

Multiply.
$$A = \frac{1}{2} \times \underline{\quad} = \underline{\quad}$$

So, the area of each tabletop is _____ m^2.

Math Talk

MATHEMATICAL PRACTICES ①

Describe Relationships Describe the relationship between the area of a trapezoid and the area of a parallelogram with the same height and a base equal to the sum of the trapezoid's bases.

🔑 Example 1 Find the area of the trapezoid.

Write the formula.

$$A = \frac{1}{2}(b_1 + b_2)h$$

Substitute 4.6 for b_1, 9.4 for b_2, and 3.5 for h.

$$A = \frac{1}{2} \times (\underline{\hspace{1cm}} + \underline{\hspace{1cm}}) \times 3.5$$

Add.

$$A = \frac{1}{2} \times \underline{\hspace{1cm}} \times 3.5$$

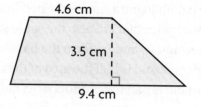

Multiply.

$$A = \underline{\hspace{1cm}} \times 3.5 = \underline{\hspace{1cm}}$$

So, the area of the trapezoid is _____ cm².

🔑 Example 2 The area of the trapezoid is 702 in.² Find the height of the trapezoid.

Write the formula.

$$A = \frac{1}{2}(b_1 + b_2)h$$

Substitute 702 for A, 20 for b_1, and 34 for b_2.

$$702 = \frac{1}{2} \times (20 + \underline{\hspace{1cm}}) \times h$$

Add within the parentheses.

$$702 = \frac{1}{2} \times \underline{\hspace{1cm}} \times h$$

Multiply $\frac{1}{2}$ and 54.

$$702 = \underline{\hspace{1cm}} \times h$$

Use the Division Property of Equality.

$$\frac{702}{\boxed{}} = \frac{\boxed{} \times h}{\boxed{}}$$

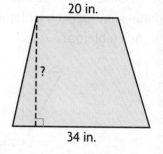

Simplify.

$$\underline{\hspace{1cm}} = h$$

So, the height of the trapezoid is _____ in.

Math Talk

MATHEMATICAL PRACTICES ①

Analyze Relationships
Explain why the formula for the area of a trapezoid contains the expression $b_1 + b_2$.

• **MATHEMATICAL PRACTICE ⑥** **Attend to Precision** Name all the trapezoids that have more than one pair of parallel sides. Do you have to use the formula for area of a trapezoid to find the area of these types of trapezoids? Explain why or why not.

_____

_____

_____

Name _____

1. Find the area of the trapezoid.

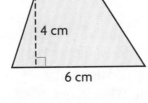
3 cm
4 cm
6 cm

$$A = \frac{1}{2}(b_1 + b_2)h$$

$$A = \frac{1}{2} \times (\underline{\hspace{1cm}} + \underline{\hspace{1cm}}) \times 4$$

$$A = \frac{1}{2} \times \underline{\hspace{1cm}} \times 4$$

$$A = \underline{\hspace{1cm}} \text{ cm}^2$$

2. The area of the trapezoid is 45 ft². Find the height of the trapezoid.

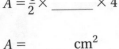
8 ft
h
10 ft

h = _____

3. Find the area of the trapezoid.

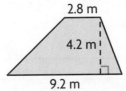
43 mm
18 mm
17 mm

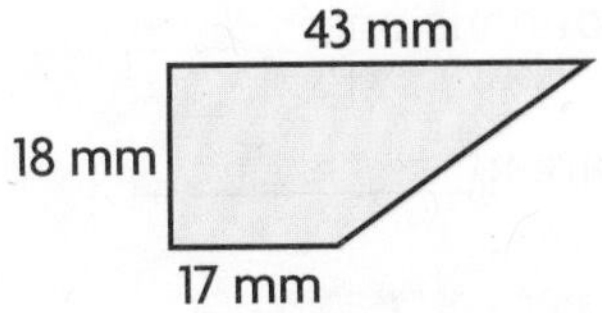

A = _____

Math Talk

MATHEMATICAL PRACTICES ①

Analyze Two trapezoids have the same bases and the same height. Are the areas equal? Must the trapezoids have the same shape?

On Your Own

Find the area of the trapezoid.

4.
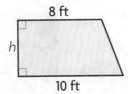
21 in.
14 in.
17 in.

A = _____

5.
2.8 m
4.2 m
9.2 m

A = _____

Find the height of the trapezoid.

6.
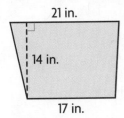
12.5 in.
Area = 500 in.²
h
27.5 in.

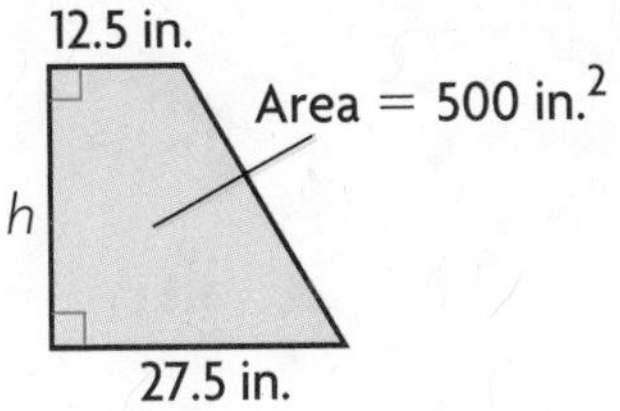

h = _____

7.
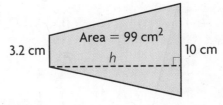
Area = 99 cm²
3.2 cm
h
10 cm

h = _____

Problem Solving · Applications Real World

Home Plate

17 in.

8.5 in.   8.5 in.

17 in.

12 in.   12 in.

Use the diagram for 8–9.

8. **GO DEEPER** A baseball home plate can be divided into two trapezoids with the dimensions shown in the drawing. Find the area of home plate.

_____

9. Suppose you cut home plate along the dotted line and rearranged the pieces to form a rectangle. What would the dimensions and the area of the rectangle be?

dimensions: _____

area: _____

10. **THINK SMARTER** A pattern used for tile floors is shown. A side of the inner square measures 10 cm, and a side of the outer square measures 30 cm. What is the area of one of the yellow trapezoid tiles?

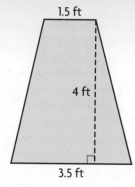

_____

WRITE ▸ Math
Show Your Work

11. **MATHEMATICAL PRACTICE ③ Verify the Reasoning of Others** A trapezoid has a height of 12 cm and bases with lengths of 14 cm and 10 cm. Tina says the area of the trapezoid is 288 cm². Find her error, and correct the error.

_____

12. **THINK SMARTER** Which expression can be used to find the area of the trapezoid? Mark all that apply.

1.5 ft

4 ft

3.5 ft

Ⓐ $\frac{1}{2} \times (4 + 1.5) \times 3.5$

Ⓑ $\frac{1}{2} \times (1.5 + 3.5) \times 4$

Ⓒ $\frac{1}{2} \times (4 + 3.5) \times 1.5$

Ⓓ $\frac{1}{2} \times (5) \times 4$

Area of Trapezoids

Common Core **COMMON CORE STANDARD—6.G.A.1**
*Solve real-world and mathematical problems
involving area, surface area, and volume.*

Find the area of the trapezoid.

1. $A = \frac{1}{2}(b_1 + b_2)h$

$A = \frac{1}{2} \times (\underline{11} + \underline{17}) \times 18$

$A = \frac{1}{2} \times \underline{28} \times 18$

$A = \underline{252}$ cm^2

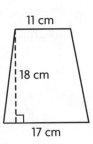

11 cm

18 cm

17 cm

2.

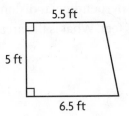

5.5 ft

5 ft

6.5 ft

$A = \underline{\hspace{3cm}}$

3.

0.2 cm

0.2 cm

0.6 cm

$A = \underline{\hspace{3cm}}$

4.

10 in.

$2\frac{1}{2}$ in.

5 in.

$A = \underline{\hspace{3cm}}$

Problem Solving · Real World

5. Sonia makes a wooden frame around a square picture. The frame is made of 4 congruent trapezoids. The shorter base is 9 in., the longer base is 12 in., and the height is 1.5 in. What is the area of the picture frame?

_____

6. Bryan cuts a piece of cardboard in the shape of a trapezoid. The area of the cutout is 43.5 square centimeters. If the bases are 6 centimeters and 8.5 centimeters long, what is the height of the trapezoid?

_____

7. **WRITE** ▸*Math* Use the formula for the area of a trapezoid to find the height of a trapezoid with bases 8 inches and 6 inches and an area of 112 square inches.

_____

Lesson Check (6.G.A.1, 6.EE.A.2c)

1. Dominic is building a bench with a seat in the shape of a trapezoid. One base is 5 feet. The other base is 4 feet. The perpendicular distance between the bases is 2.5 feet. What is the area of the seat?

2. Molly is making a sign in the shape of a trapezoid. One base is 18 inches and the other is 30 inches. How high must she make the sign so its area is 504 square inches?

Spiral Review (6.NS.C.6c, 6.RP.A.3d, 6.EE.A.2c)

3. Write these numbers in order from least to greatest.

   $3\frac{3}{10}$ 3.1 $3\frac{1}{4}$

4. Write these lengths in order from least to greatest.

   2 yards   5.5 feet   70 inches

5. To find the cost for a group to enter the museum, the ticket seller uses the expression $8a + 3c$ in which a represents the number of adults and c represents the number of children in the group. How much should she charge a group of 3 adults and 5 children?

6. Brian frosted a cake top shaped like a parallelogram with a base of 13 inches and a height of 9 inches. Nancy frosted a triangular cake top with a base of 15 inches and a height of 12 inches. Which cake's top had the greater area? How much greater was it?

FOR MORE PRACTICE
GO TO THE
Personal Math Trainer

Name _____

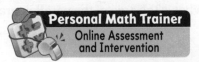

Personal Math Trainer
Online Assessment
and Intervention

Vocabulary

Choose the best term from the box to complete the sentence.

Vocabulary

area

congruent

parallelogram

trapezoid

1. A _____ is a quadrilateral that always has two pairs of parallel sides. (p. 533)

2. The measure of the number of unit squares needed to cover a surface

   without any gaps or overlaps is called the _____. (p. 533)

3. Figures with the same size and shape are _____. (p. 539)

Concepts and Skills

Find the area. (6.G.A.1, 6.EE.A.2c)

4.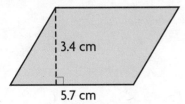

   3.4 cm

   5.7 cm

   _____

5.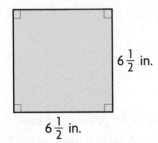

   $6\frac{1}{2}$ in.

   $6\frac{1}{2}$ in.

   _____

6.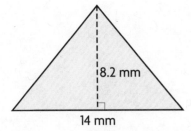

   8.2 mm

   14 mm

   _____

7.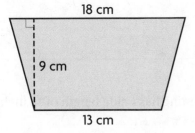

   18 cm

   9 cm

   13 cm

   _____

8. A parallelogram has an area of 276 square meters and a base measuring 12 meters. What is the height of the parallelogram?

   _____

9. The base of a triangle measures 8 inches and the area is 136 square inches. What is the height of the triangle?

   _____

10. The height of a parallelogram is 3 times the base. The base measures 4.5 cm. What is the area of the parallelogram? (6.G.A.1)

_____

11. A triangular window pane has a base of 30 inches and a height of 24 inches. What is the area of the window pane? (6.G.A.1)

_____

12. The courtyard behind Jennie's house is shaped like a trapezoid. The bases measure 8 meters and 11 meters. The height of the trapezoid is 12 meters. What is the area of the courtyard? (6.G.A.1)

_____

13. 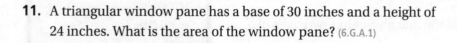 Rugs sell for $8 per square foot. Beth bought a 9-foot-long rectangular rug for $432. How wide was the rug? (6.G.A.1, 6.EE.A.2c)

_____

14. A square painting has a side length of 18 inches. What is the area of the painting? (6.G.A.1, 6.EE.A.2c)

_____

Area of Regular Polygons

Essential Question How can you find the area of regular polygons?

 Common Core · Geometry—6.G.A.1
Also 6.EE.A.2C
MATHEMATICAL PRACTICES
MP1, MP6, MP8

🔑 Unlock the Problem Real World

Emory is making a patch for his soccer ball. The patch he is using is a regular polygon. A **regular polygon** is a polygon in which all sides have the same length and all angles have the same measure. Emory needs to find the area of a piece of material shaped like a regular pentagon.

🔓 Activity

You can find the area of a regular polygon by dividing the polygon into congruent triangles.

- Draw line segments from each vertex to the center of the pentagon to divide it into five congruent triangles.

- You can find the area of one of the triangles if you know the side length of the polygon and the height of the triangle.

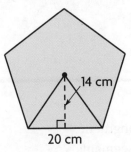

14 cm

20 cm

Math Talk

MATHEMATICAL PRACTICES ⑥

Explain How do you determine the number of congruent triangles a regular polygon should be divided into in order to find the area?

- Find the area of one triangle.

  Write the formula. $\qquad\qquad\qquad A = \frac{1}{2} bh$

  Substitute 20 for *b* and 14 for *h*. $\qquad A = \frac{1}{2} \times$ _____ $\times$ _____

  Simplify. $\qquad\qquad\qquad\qquad A =$ _____ cm²

- Find the area of the regular polygon by multiplying the number of triangles by the area of one triangle.

  $A =$ _____ $\times$ _____ $=$ _____ cm²

So, the area of the pentagon-shaped piece is _____ .

🔑 Example   Find the area of the regular polygon.

STEP 1 Draw line segments from each vertex to the center of the hexagon.

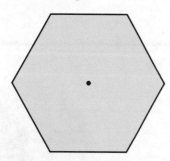

Into how many congruent triangles did you divide the figure? _____

STEP 2 Find the area of one triangle.

Write the formula.

$$A = \frac{1}{2}bh$$

Substitute 4.2 for b and 3.6 for h.

$$A = \frac{1}{2} \times \underline{\hspace{1cm}} \times \underline{\hspace{1cm}}$$

Simplify.

$$A = \underline{\hspace{1cm}} \text{ m}^2$$

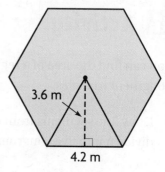

3.6 m

4.2 m

STEP 3 Find the area of the hexagon.

$$A = \underline{\hspace{1cm}} \times \underline{\hspace{1cm}} = \underline{\hspace{2cm}} \text{ m}^2$$

So, the area of the hexagon is _____ m^2

---

1. **MATHEMATICAL PRACTICE 8** **Use Repeated Reasoning** Into how many congruent triangles can you divide a regular decagon by drawing line segments from each vertex to the center of the decagon? Explain.

_____

_____

2. **THINK SMARTER** In an *irregular polygon*, the sides do not all have the same length and the angles do not all have the same measure. Could you find the area of an irregular polygon using the method you used in this lesson? Explain your reasoning.

_____

_____

Name _____

Find the area of the regular polygon.

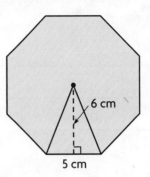

1. number of congruent triangles inside the figure: _____

   area of each triangle: $\frac{1}{2} \times$ _____ $\times$ _____ = _____ cm²

   area of octagon: _____ $\times$ _____ = _____ cm²

6 cm

5 cm

☑2.

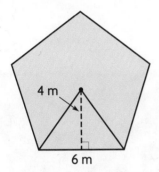

4 m

6 m

☑3.

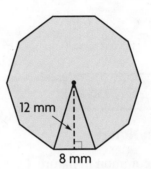

12 mm

8 mm

Math Talk

MATHEMATICAL PRACTICES ①

Describe the information you must have about a regular polygon in order to find its area.

On Your Own

Find the area of the regular polygon.

4.

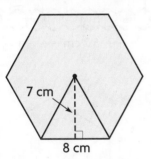

7 cm

8 cm

5.

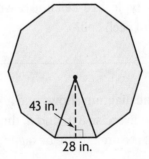

43 in.

28 in.

6. **MATHEMATICAL PRACTICE ⑥** **Explain** A regular pentagon is divided into congruent triangles by drawing a line segment from each vertex to the center. Each triangle has an area of 24 cm². Explain how to find the area of the pentagon.

_____

_____

7. THINK SMARTER Name the polygon and find its area.
Show your work.

_____

_____

_____

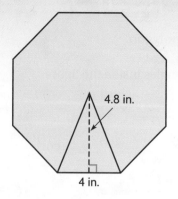

4.8 in.

4 in.

Regular Polygons in Nature

Regular polygons are common in nature. One of the best-known examples of regular polygons in nature is the small hexagonal cells in honeycombs constructed by honeybees. The cells are where bee larvae grow. Honeybees store honey and pollen in the hexagonal cells. Scientists can measure the health of a bee population by the size of the cells.

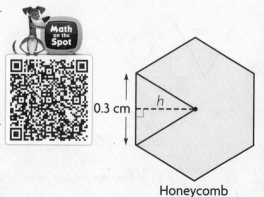

8. Cells in a honeycomb vary in width. To find the average width of a cell, scientists measure the combined width of 10 cells, and then divide by 10.

The figure shows a typical 10-cell line of worker bee cells. What is the width of each cell?

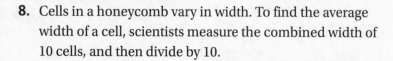

← 5.2 cm →

_____

9. THINK SMARTER The diagram shows one honeycomb cell. Use your answer to Exercise 8 to find h, the height of the triangle. Then find the area of the hexagonal cell.

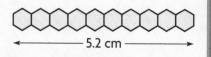

0.3 cm h

Honeycomb

_____

10. GO DEEPER A rectangular honeycomb measures 35.1 cm by 32.4 cm. Approximately how many cells does it contain?

_____

Name _____

Area of Regular Polygons

COMMON CORE STANDARD—6.G.A.1
Solve real-world and mathematical problems involving area, surface area, and volume.

Find the area of the regular polygon.

1.

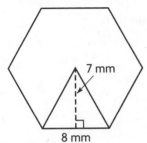

7 mm

8 mm

number of congruent triangles inside the figure: ___6___

area of each triangle: $\frac{1}{2} \times$ ___8___ $\times$ ___7___ = ___28___ mm^2

area of hexagon: ___168 mm^2___

2.

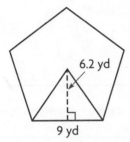

6.2 yd

9 yd

_____

3.

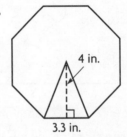

4 in.

3.3 in.

_____

Problem Solving · Real World

4. Stu is making a stained glass window in the shape of a regular pentagon. The pentagon can be divided into congruent triangles, each with a base of 8.7 inches and a height of 6 inches. What is the area of the window?

_____

5. A dinner platter is in the shape of a regular decagon. The platter has an area of 161 square inches and a side length of 4.6 inches. What is the area of each triangle? What is the height of each triangle?

_____

6. **WRITE** ▸*Math* A square has sides that measure 6 inches. Explain how to use the method in this lesson to find the area of the square.

_____

_____

_____

Lesson Check (6.G.A.1, 6.EE.A.2c)

1. What is the area of the regular hexagon?

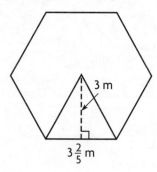

3 m

$3\frac{2}{5}$ m

2. A regular 7-sided figure is divided into 7 congruent triangles, each with a base of 12 inches and a height of 12.5 inches. What is the area of the 7-sided figure?

Spiral Review (6.EE.A.2c, 6.EE.B.5, 6.EE.C.9, 6.G.A.1)

3. Which inequalities have $b = 4$ as one of its solutions?

$2 + b \geq 2$ $3b \leq 14$

$8 - b \leq 15$ $b - 3 \geq 5$

4. Each song that Tara downloads costs $1.25. She graphs the relationship that gives the cost y in dollars of downloading x songs. Name one ordered pair that is a point on the graph of the relationship.

5. What is the area of triangle ABC?

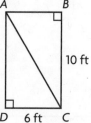

A   B

10 ft

D   6 ft   C

6. Marcia cut a trapezoid out of a large piece of felt. The trapezoid has a height of 9 cm and bases of 6 cm and 11 cm. What is the area of Marcia's felt trapezoid?

FOR MORE PRACTICE GO TO THE Personal Math Trainer

Name _____

Composite Figures

Essential Question How can you find the area of composite figures?

Common Core Geometry—6.G.A.1
Also 6.EE.A.2c
MATHEMATICAL PRACTICES
MP1, MP6

A **composite figure** is made up of two or more simpler figures, such as triangles and quadrilaterals.

? Unlock the Problem Real World

The new entryway to the fun house at Happy World Amusement Park is made from the shapes shown in the diagram. It will be painted bright green. Juanita needs to know the area of the entryway to determine how much paint to buy. What is the area of the entryway?

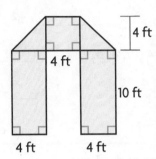

Find the area of the entryway.

STEP 1 Find the area of the rectangles.

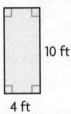

Write the formula.	$A = lw$
Substitute the values for l and w and evaluate.	$A = 10 \times$ _____ = _____
Find the total area of two rectangles.	$2 \times$ _____ = _____ ft^2

STEP 2 Find the area of the triangles.

Write the formula.	$A = \frac{1}{2}bh$
Substitute the values for b and h and evaluate.	$A = \frac{1}{2} \times 4 \times$ _____ = _____
Find the total area of two triangles.	$2 \times$ _____ = _____ ft^2

STEP 3 Find the area of the square.

| Write the formula. | $A = s^2$ |
| Substitute the value for s. | $A = ($ _____ $)^2 =$ _____ ft^2 |

STEP 4 Find the total area of the composite figure.

Add the areas. $A = 80\ ft^2 +$ _____ $ft^2 +$ _____ $ft^2 =$ _____ ft^2

So, Juanita needs to buy enough paint to cover _____ ft^2.

Math Talk MATHEMATICAL PRACTICES ①

Describe other ways you could divide up the composite figure.

Example 1   Find the area of the composite figure shown.

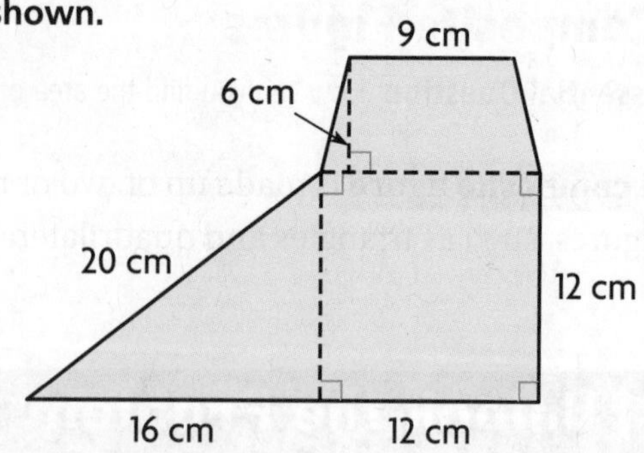

STEP 1 Find the area of the triangle, the square, and the trapezoid.

9 cm

6 cm

20 cm

12 cm

16 cm   12 cm

area of triangle $A = \frac{1}{2}bh = \frac{1}{2} \times 16 \times \underline{\hspace{1cm}}$

$= \underline{\hspace{1cm}} \text{ cm}^2$

area of square $A = s^2 = (\underline{\hspace{1cm}})^2$

$= \underline{\hspace{2cm}} \text{ cm}^2$

area of trapezoid $A = \frac{1}{2}(b_1 + b_2)h = \frac{1}{2} \times (\underline{\hspace{1cm}} + \underline{\hspace{1cm}}) \times \underline{\hspace{1cm}}$

$= \frac{1}{2} \times \underline{\hspace{1cm}} \times 6$

$= \underline{\hspace{1cm}} \text{ cm}^2$

STEP 2 Find the total area of the figure.

total area $A = \underline{\hspace{1cm}} \text{ cm}^2 + \underline{\hspace{1cm}} \text{ cm}^2 + \underline{\hspace{1cm}} \text{ cm}^2$

$= \underline{\hspace{1cm}} \text{ cm}^2$

So, the area of the figure is _____ cm².

Example 2   Find the area of the shaded region.

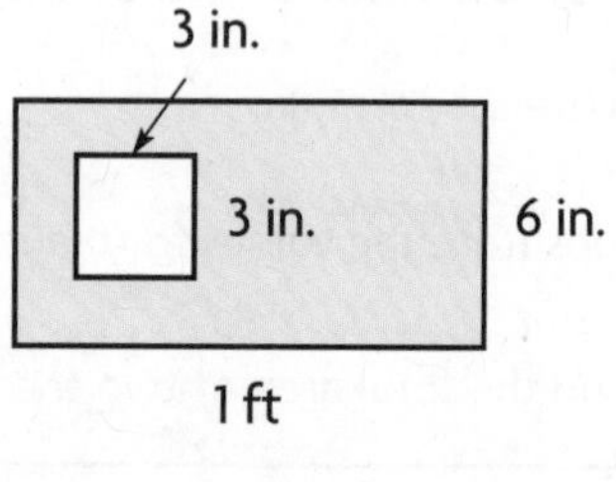

3 in.

3 in.   6 in.

1 ft

STEP 1 Find the area of the rectangle and the square.

area of rectangle (1 ft = 12 in.) $A = lw = \underline{\hspace{1cm}} \times \underline{\hspace{1cm}}$

$A = \underline{\hspace{1cm}} \text{ in.}^2$

area of square $A = s^2 = (\underline{\hspace{1cm}})^2$

$A = \underline{\hspace{1cm}} \text{ in.}^2$

STEP 2 Subtract the area of the square from the area of the rectangle.

area of shaded region $A = \underline{\hspace{1cm}} \text{ in.}^2 - \underline{\hspace{1cm}} \text{ in.}^2$

$A = \underline{\hspace{1cm}} \text{ in.}^2$

So, the area of the shaded region is _____ in.²

Name _____

Share and Show

1. Find the area of the figure.

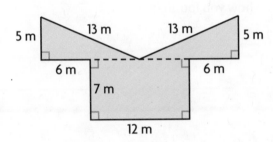

4 ft
3 ft
10 ft
5 ft   5 ft

area of one rectangle

$A = lw$

$A =$ _____ × _____ = _____ ft²

area of two rectangles

$A = 2 ×$ _____ = _____ ft²

length of base of triangle

$b =$ _____ ft + _____ ft + _____ ft

= _____ ft

area of triangle

$A = \frac{1}{2} bh$

$A = \frac{1}{2} ×$ _____ × _____ = _____ ft²

area of composite figure

$A =$ _____ ft² + _____ ft² = _____ ft²

Find the area of the figure.

2.

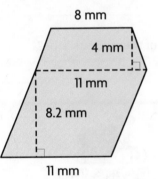

8 mm
4 mm
11 mm
8.2 mm
11 mm

3.

5 m   13 m   13 m   5 m
6 m   6 m
7 m
12 m

Math Talk

MATHEMATICAL PRACTICES ⑥

Explain how to find the area of a composite figure.

On Your Own

4. Find the area of the figure.

8 in.
10 in.
6 in.
8 in.   16 in.

5. MATHEMATICAL PRACTICE ⑥ **Attend to Precision** Find the area of the shaded region.

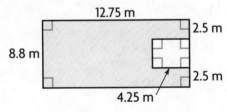

12.75 m
2.5 m
8.8 m
2.5 m
4.25 m

© Houghton Mifflin Harcourt Publishing Company

Unlock the Problem

6. **GO DEEPER** Marco made the banner shown at the right. What is the area of the yellow shape?

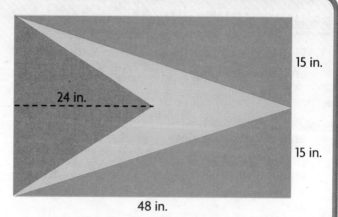

15 in.

24 in.

15 in.

48 in.

a. Explain how you could find the area of the yellow shape if you knew the areas of the green and red shapes and the area of the entire banner.

_____

_____

_____

c. What is the area of the red shape? What is the area of each green shape?

_____

b. What is the area of the entire banner? Explain how you found it.

_____

_____

d. What equation can you write to find A, the area of the yellow shape?

_____

e. What is the area of the yellow shape?

_____

7. There are 6 rectangular flower gardens each measuring 18 feet by 15 feet in a rectangular city park measuring 80 feet by 150 feet. How many square feet of the park are not used for flower gardens?

_____

Personal Math Trainer

8. **THINK SMARTER +** Sabrina wants to replace the carpet in a few rooms of her house. Select the expression she can use to find the total area of the floor that will be covered. Mark all that apply.

(A) $8 \times 22 + 130 + \frac{1}{2} \times 10 \times 9$

(B) $18 \times 22 - \frac{1}{2} \times 10 \times 9$

(C) $18 \times 13 + \frac{1}{2} \times 10 \times 9$

(D) $\frac{1}{2} \times (18 + 8) \times 22$

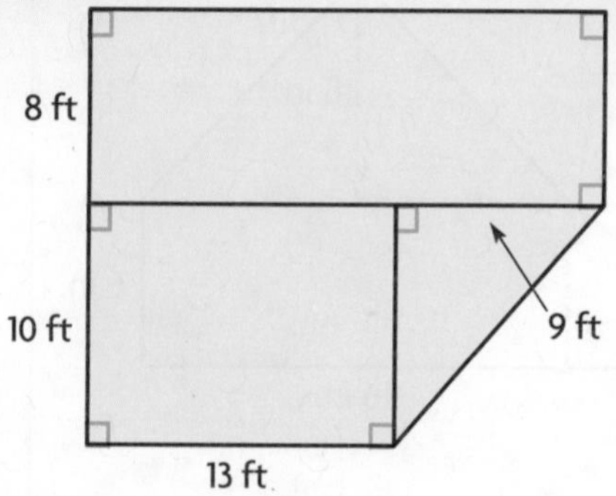

8 ft

10 ft

9 ft

13 ft

Composite Figures

Find the area of the figure.

Common Core
COMMON CORE STANDARD—6.G.A.1
Solve real-world and mathematical problems involving area, surface area, and volume.

1.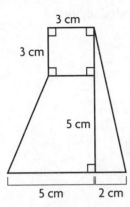

area of square

$A = s \times s$

$= \underline{\quad 3 \quad} \times \underline{\quad 3 \quad} = \underline{\quad 9 \quad}$ cm²

area of triangle

$A = \frac{1}{2}bh$

$= \frac{1}{2} \times \underline{\quad 2 \quad} \times \underline{\quad 8 \quad} = \underline{\quad 8 \quad}$ cm²

area of trapezoid

$A = \frac{1}{2}(b_1 + b_2)h$

$= \frac{1}{2} \times (\underline{\quad 5 \quad} + \underline{\quad 3 \quad}) \times \underline{\quad 5 \quad} = \underline{\quad 20 \quad}$ cm²

area of composite figure

$A = \underline{\quad 9 \quad}$ cm² $+ \underline{\quad 8 \quad}$ cm² $+ \underline{\quad 20 \quad}$ cm²

$= \underline{\quad 37 \quad}$ cm²

2.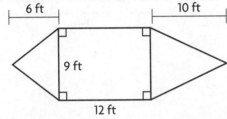

_____

3.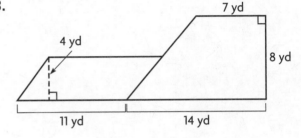

_____

Problem Solving · Real World

4. Janelle is making a poster. She cuts a triangle out of poster board. What is the area of the poster board that she has left?

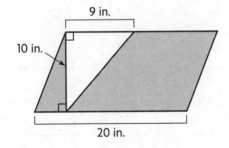

_____

5. Michael wants to place grass on the sides of his lap pool. Find the area of the shaded regions that he wants to cover with grass.

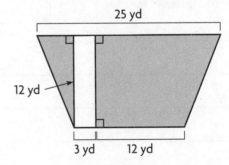

_____

6. **WRITE** ▸*Math* Describe one or more situations in which you need to subtract to find the area of a composite figure.

_____

_____

Lesson Check (6.G.A.1, 6.EE.A.2c)

1. What is the area of the composite figure?

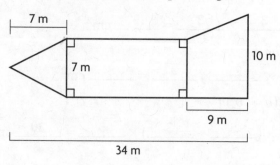

2. What is the area of the shaded region?

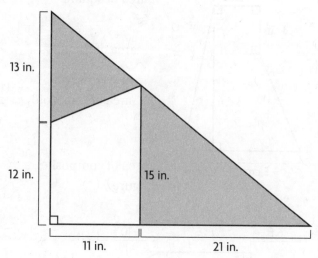

_____

_____

Spiral Review (6.EE.A.2c, 6.EE.B.8, 6.EE.C.9, 6.G.A.1)

3. In Maritza's family, everyone's height is greater than 60 inches. Write an inequality that represents the height h, in inches, of any member of Maritza's family.

_____

_____

_____

_____

4. The linear equation $y = 2x$ represents the cost y for x pounds of apples. Which ordered pair lies on the graph of the equation?

_____

_____

_____

_____

5. Two congruent triangles fit together to form a parallelogram with a base of 14 inches and a height of 10 inches. What is the area of each triangle?

6. A regular hexagon has sides measuring 7 inches. If the hexagon is divided into 6 congruent triangles, each has a height of about 6 inches. What is the approximate area of the hexagon?

_____

_____

FOR MORE PRACTICE GO TO THE
Personal Math Trainer

Problem Solving • Changing Dimensions

Essential Question How can you use the strategy *find a pattern* to show how changing dimensions affects area?

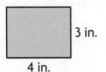

Geometry—
6.G.A.1

MATHEMATICAL PRACTICES
MP2, MP4, MP8

 Unlock the Problem *Real World*

Jason has created a 3-in. by 4-in. rectangular design to be made into mouse pads. To manufacture the pads, the dimensions will be multiplied by 2 or 3. How will the area of the design be affected?

Use the graphic organizer to help you solve the problem.

3 in.

4 in.

Read the Problem

What do I need to find?	What information do I need to use?	How will I use the information?
I need to find how _____ will be affected by changing the _____.	I need to use _____ of the original design and _____ _____ _____.	I can draw a sketch of each rectangle and calculate _____ of each. Then I can look for _____ in my results.

Solve the Problem

Sketch	Dimensions	Multiplier	Area
	3 in. by 4 in.	1	$A = 3 \times 4 = 12$ in.²
6 in. / 8 in.	6 in. by 8 in.	2	$A =$ _____ $\times$ _____ $=$ _____ in.²
9 in. / 12 in.			

So, when the dimensions are multiplied by 2, the area is

multiplied by _____. When the dimensions are multiplied

by 3, the area is multiplied by _____.

 Math Talk

MATHEMATICAL PRACTICES ②

Reasoning What would happen to the area of a rectangle if the dimensions were multiplied by 4?

🔑 Try Another Problem

A stained-glass designer is reducing the dimensions of an earlier design. The dimensions of the triangle shown will be multiplied by $\frac{1}{2}$ or $\frac{1}{4}$. How will the area of the design be affected? Use the graphic organizer to help you solve the problem.

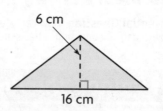

6 cm

16 cm

Read the Problem

What do I need to find?	What information do I need to use?	How will I use the information?

Solve the Problem

Sketch	Multiplier	Area
	1	$A = \frac{1}{2} \times 16 \times$ _____ = _____ cm²
3 cm 8 cm	$\frac{1}{2}$	

So, when the dimensions are multiplied by $\frac{1}{2}$, the area is multiplied by

_____. When the dimensions are multiplied by _____, the area is

multiplied by _____.

MATHEMATICAL PRACTICES ⑧

Generalize What happens to the area of a triangle when the dimensions are multiplied by a number *n*?

Name _____

Unlock the Problem

✓ Plan your solution by deciding on the steps you will use.

✓ Find the original area and the new area, and then compare the two.

✓ Look for patterns in your results.

Share and Show

1. The dimensions of a 2-cm by 6-cm rectangle are multiplied by 5. How is the area of the rectangle affected?

First, find the original area:

_____

Next, find the new area:

_____

So, the area is multiplied by _____.

⬚ **WRITE** ⟩*Math* • **Show Your Work** • • •

2. **THINK SMARTER** What if the dimensions of the original rectangle in Exercise 1 had been multiplied by $\frac{1}{2}$? How would the area have been affected?

_____

_____

_____

_____

3. Evan bought two square rugs. The larger one measured 12 ft square. The smaller one had an area equal to $\frac{1}{4}$ the area of the larger one. What fraction of the side lengths of the larger rug were the side lengths of the smaller one?

_____

4. **GO DEEPER** On Silver Island, a palm tree, a giant rock, and a buried treasure form a triangle with a base of 100 yd and a height of 50 yd. On a map of the island, the three landmarks form a triangle with a base of 2 ft and a height of 1 ft. How many times the area of the triangle on the map is the area of the actual triangle?

_____

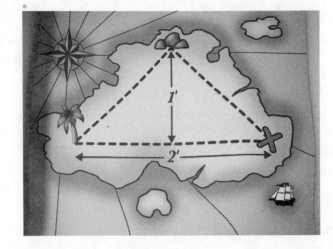

On Your Own

5. A square game board is divided into smaller squares, each with sides one-ninth the length of the sides of the board. Into how many squares is the game board divided?

_____

6. **THINK SMARTER** Flynn County is a rectangle measuring 9 mi by 12 mi. Gibson County is a rectangle with an area 6 times the area of Flynn County and a width of 16 mi. What is the length of Gibson County?

_____

7. **MATHEMATICAL PRACTICE 4** Use Diagrams Carmen left her house and drove 10 mi north, 15 mi east, 13 mi south, 11 mi west, and 3 mi north. How far was she from home?

_____

8. **GO DEEPER** Bernie drove from his house to his cousin's house in 6 hours at an average rate of 52 mi per hr. He drove home at an average rate of 60 mi per hr. How long did it take him to drive home?

_____

Personal Math Trainer

9. **THINK SMARTER +** Sophia wants to enlarge a 5-inch by 7-inch rectangular photo by multiplying the dimensions by 3.

Find the area of the original photo and the enlarged photo. Then explain how the area of the original photo is affected.

_____

_____

_____

Problem Solving • Changing Dimensions

Common Core **COMMON CORE STANDARD—6.G.A.1**
Solve real-world and mathematical problems involving area, surface area, and volume.

Read each problem and solve.

1. The dimensions of a 5-in. by 3-in. rectangle are multiplied by 6. How is the area affected?

   $$l = 6 \times 5 = 30 \text{ in.}$$

   new dimensions: $w = 6 \times 3 = 18$ in.

   original area: $A = 5 \times 3 = 15$ in.^{2}

   new area: $A = 30 \times 18 = 540$ in.^{2}

   $\dfrac{\text{new area}}{\text{original area}} = \dfrac{540}{15} = 36$

   The area was multiplied by ____36____.

2. The dimensions of a 7-cm by 2-cm rectangle are multiplied by 3. How is the area affected?

   multiplied by _____

3. The dimensions of a 3-ft by 6-ft rectangle are multiplied by $\frac{1}{3}$. How is the area affected?

   multiplied by _____

4. The dimensions of a triangle with base 10 in. and height 4.8 in. are multiplied by 4. How is the area affected?

   multiplied by _____

5. The dimensions of a 1-yd by 9-yd rectangle are multiplied by 5. How is the area affected?

   multiplied by _____

6. The dimensions of a 4-in. square are multiplied by 3. How is the area affected?

   multiplied by _____

7. The dimensions of a triangle are multiplied by $\frac{1}{4}$. The area of the smaller triangle can be found by multiplying the area of the original triangle by what number?

   _____

8. **WRITE** ▸*Math* Write and solve a word problem that involves changing the dimensions of a figure and finding its area.

   _____

   _____

Lesson Check (6.G.A.1)

1. The dimensions of Rectangle A are 6 times the dimensions of Rectangle B. How do the areas of the rectangles compare?

_____

_____

_____

2. A model of a triangular piece of jewelry has an area that is $\frac{1}{4}$ the area of the jewelry. How do the dimensions of the triangles compare?

_____

_____

_____

Spiral Review (6.RP.A.3c, 6.EE.A.2c, 6.EE.B.8, 6.G.A.1)

3. Gina made a rectangular quilt that was 5 feet wide and 6 feet long. She used yellow fabric for 30% of the quilt. What was the area of the yellow fabric?

_____

4. Graph $y > 3$ on a number line.

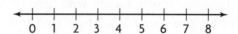

5. The parallelogram below is made from two congruent trapezoids. What is the area of the shaded trapezoid?

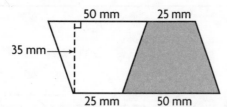

6. A rectangle has a length of 24 inches and a width of 36 inches. A square with side length 5 inches is cut from the middle and removed. What is the area of the figure that remains?

_____

_____

FOR MORE PRACTICE
GO TO THE
Personal Math Trainer

Figures on the Coordinate Plane

Essential Question How can you plot polygons on a coordinate plane and find their side lengths?

Common Core   Geometry—6.G.A.3
Also 6.NS.C.8
MATHEMATICAL PRACTICES
MP4, MP6, MP8

Unlock the Problem Real World

The world's largest book is a collection of photographs from the Asian nation of Bhutan. A book collector models the rectangular shape of the open book on a coordinate plane. Each unit of the coordinate plane represents one foot. The book collector plots the vertices of the rectangle at $A(9, 3)$, $B(2, 3)$, $C(2, 8)$, and $D(9, 8)$. What are the dimensions of the open book?

- What two dimensions do you need to find?

_____

_____

Plot the vertices and find the dimensions of the rectangle.

STEP 1 Complete the rectangle on the coordinate plane.

Plot points $C(2, 8)$ and $D(9, 8)$.
Connect the points to form a rectangle.

_____

STEP 2 Find the length of the rectangle.

Find the distance between points $A(9, 3)$ and $B(2, 3)$.

The y-coordinates are the same, so the points lie on a _____ line.

Think of the horizontal line passing through A and B as a number line.

Horizontal distance of A from 0: $|9| =$ _____ ft

Horizontal distance of B from 0: $|2| =$ _____ ft

Subtract to find the distance from A to B: _____ − _____ = _____ ft.

STEP 3 Find the width of the rectangle.

Find the distance between points $C(2, 8)$ and $B(2, 3)$.

The x-coordinates are the same, so the points lie on a _____ line.

Think of the vertical line passing through C and B as a number line.

Vertical distance of C from 0: $|8| =$ _____ ft

Vertical distance of B from 0: $|3| =$ _____ ft

Subtract to find the distance from C to B: _____ − _____ = _____ ft.

So, the dimensions of the open book are _____ ft by _____ ft.

Math Talk   MATHEMATICAL PRACTICES ⑥

Explain How do you know whether to add or subtract the absolute values to find the distance between the vertices of the rectangle?

CONNECT You can use properties of quadrilaterals to help you find unknown vertices. The properties can also help you graph quadrilaterals on the coordinate plane.

🔑 Example Find the unknown vertex, and then graph.

Three vertices of parallelogram *PQRS* are *P*(4, 2), *Q*(3, ⁻3), and *R*(⁻3, ⁻3). Give the coordinates of vertex *S* and graph the parallelogram.

STEP 1

Plot the given points on the coordinate plane.

STEP 2

The opposite sides of a parallelogram are _____.

They have the same _____.

Since the length of side $\overline{RQ}$ is _____ units, the length of

side _____ must also be _____ units.

STEP 3

Start at point *P*. Move horizontally _____ units to the

_____ to find the location of the remaining
vertex, *S*. Plot a point at this location.

STEP 4

Draw the parallelogram. Check that opposite sides are parallel and congruent.

So, the coordinates of the vertex *S* are _____.

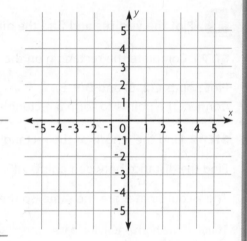

1. **MATHEMATICAL PRACTICE ⑥ Attend to Precision** Explain why vertex *S* must be to the left of vertex *P* rather than to the right of vertex *P*.

_____

_____

2. Describe how you could find the area of parallelogram *PQRS* in square units.

_____

_____

Name _____

 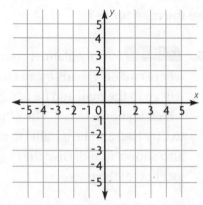
1. The vertices of triangle ABC are $A(^-1, 3)$, $B(^-4, ^-2)$, and $C(2, ^-2)$. Graph the triangle and find the length of side $\overline{BC}$.

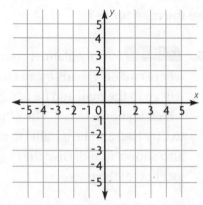

Horizontal distance of B from 0: $|^-4| = $ _____ units

Horizontal distance of C from 0: $|2| = $ _____ units

The points are in different quadrants, so add to find the

distance from B to C: _____ + _____ = _____ units.

Give the coordinates of the unknown vertex of rectangle JKLM, and graph.

2.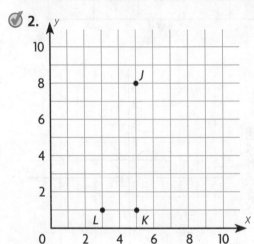

3.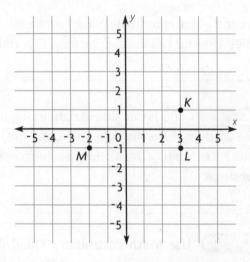

_____

4. Give the coordinates of the unknown vertex of rectangle *PQRS*, and graph.

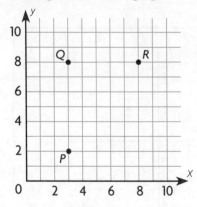

5. The vertices of pentagon *PQRST* are $P(9, 7)$, $Q(9, 3)$, $R(3, 3)$, $S(3, 7)$, and $T(6, 9)$. Graph the pentagon and find the length of side $\overline{PQ}$.

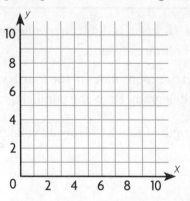

_____

© Houghton Mifflin Harcourt Publishing Company

Problem Solving · Applications

The map shows the location of some city landmarks. Use the map for 6–7.

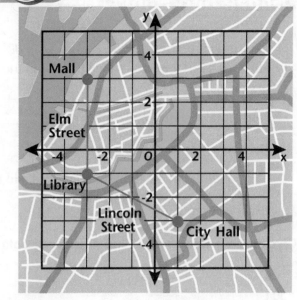

6. **GO DEEPER** A city planner wants to locate a park where two new roads meet. One of the new roads will go to the mall and be parallel to Lincoln Street which is shown in red. The other new road will go to City Hall and be parallel to Elm Street which is also shown in red. Give the coordinates for the location of the park.

_____

7. Each unit of the coordinate plane represents 2 miles. How far will the park be from City Hall?

_____

8. **THINK SMARTER** $\overline{PQ}$ is one side of right triangle PQR. In the triangle, $\angle P$ is the right angle, and the length of side $\overline{PR}$ is 3 units. Give all the possible coordinates for vertex R.

_____

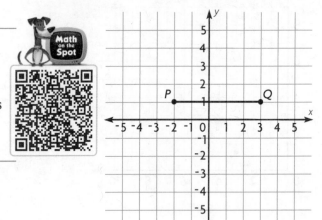

9. **MATHEMATICAL PRACTICE 6** Use Math Vocabulary Quadrilateral $WXYZ$ has vertices with coordinates $W(^-4, 0)$, $X(^-2, 3)$, $Y(2, 3)$, and $Z(2, 0)$. Classify the quadrilateral using the most exact name possible and explain your answer.

_____

_____

10. **THINK SMARTER** Kareem is drawing parallelogram $ABCD$ on the coordinate plane.

Find and label the coordinates of the fourth vertex, D, of the parallelogram. Draw the parallelogram.

What is the length of side CD? How do you know?

_____

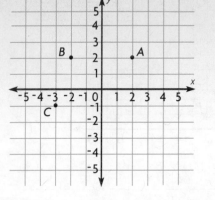

Figures on the Coordinate Plane

Common Core **COMMON CORE STANDARD—6.G.A.3**
Solve real-world and mathematical problems involving area, surface area, and volume.

1. The vertices of triangle *DEF* are $D(^-2, 3)$, $E(3, ^-2)$, and $F(^-2, ^-2)$. Graph the triangle, and find the length of side $\overline{DF}$.

   Vertical distance of *D* from 0: $|3| =$ ___**3**___ units

   Vertical distance of *F* from 0: $|^-2| =$ ___**2**___ units

   The points are in different quadrants, so add to find the

   distance from *D* to *F*: ___**3**___ + ___**2**___ = ___**5**___ units.

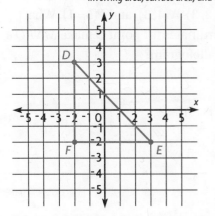

Graph the figure and find the length of side $\overline{BC}$.

2. $A(1, 4)$, $B(1, ^-2)$, $C(^-3, ^-2)$, $D(^-3, 3)$

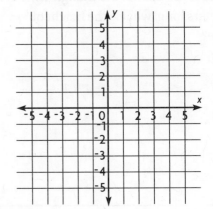

   Length of $\overline{BC} =$ _____ units

3. $A(^-1, 4)$, $B(5, 4)$, $C(5, 1)$, $D(^-1, 1)$

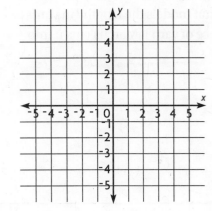

   Length of $\overline{BC} =$ _____ units

Problem Solving

4. On a map, a city block is a square with three of its vertices at $(^-4, 1)$, $(1, 1)$, and $(1, ^-4)$. What are the coordinates of the remaining vertex?

   _____

5. A carpenter is making a shelf in the shape of a parallelogram. She begins by drawing parallelogram *RSTU* on a coordinate plane with vertices $R(1, 0)$, $S(^-3, 0)$, and $T(^-2, 3)$. What are the coordinates of vertex *U*?

   _____

6. **WRITE** *Math* Explain how you would find the fourth vertex of a rectangle with vertices at $(2, 6)$, $(^-1, 4)$, and $(^-1, 6)$.

_____

_____

Lesson Check (6.G.A.3)

1. The coordinates of points M, N, and P are $M(^-2, 3)$, $N(4, 3)$, and $P(5, ^-1)$. What coordinates for point Q make $MNPQ$ a parallelogram?

_____

2. Dirk draws quadrilateral $RSTU$ with vertices $R(^-1, 2)$, $S(4, 2)$, $T(5, ^-1)$, and $U(^-2, ^-1)$. Which is the best way to classify the quadrilateral?

_____

Spiral Review (6.EE.A.2c, 6.EE.C.9, 6.G.A.1)

3. Marcus needs to cut a 5-yard length of yarn into equal pieces for his art project. Write an equation that models the length l in yards of each piece of yarn if Marcus cuts it into p pieces.

_____

4. The area of a triangular flag is 330 square centimeters. If the base of the triangle is 30 centimeters long, what is the height of the triangle?

_____

5. A trapezoid is $6\frac{1}{2}$ feet tall. Its bases are 9.2 feet and 8 feet long. What is the area of the trapezoid?

_____

6. The dimensions of the rectangle below will be multiplied by 3. How will the area be affected?

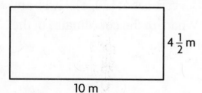

$4\frac{1}{2}$ m

10 m

_____

**FOR MORE PRACTICE
GO TO THE
Personal Math Trainer**

Name _____

1. Find the area of the parallelogram.

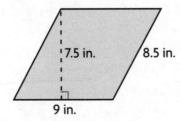

7.5 in.   8.5 in.

9 in.

The area is _____ in.²

2. A wall tile is two different colors. What is the area of the white part of the tile? Explain how you found your answer.

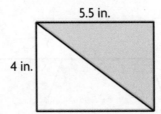

5.5 in.

4 in.

3. The area of a triangle is 36 ft². For numbers 3a–3d, select Yes or No to tell if the dimensions could be the height and base of the triangle.

3a. $h = 3$ ft, $b = 12$ ft   ○ Yes   ○ No

3b. $h = 3$ ft, $b = 24$ ft   ○ Yes   ○ No

3c. $h = 4$ ft, $b = 18$ ft   ○ Yes   ○ No

3d. $h = 4$ ft, $b = 9$ ft   ○ Yes   ○ No

4. Mario traced this trapezoid. Then he cut it out and arranged the trapezoids to form a rectangle. What is the area of the rectangle?

_____ in.²

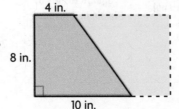

4 in.

8 in.

10 in.

Assessment Options
Chapter Test

5. The area of the triangle is 24 ft². Use the numbers to label the height and base of the triangle.

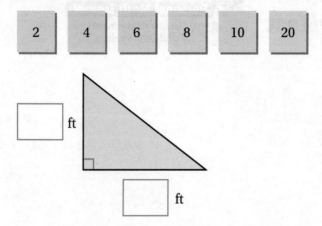

| 2 | 4 | 6 | 8 | 10 | 20 |

☐ ft

☐ ft

6. A rectangle has an area of 50 cm². The dimensions of the rectangle are multiplied to form a new rectangle with an area of 200 cm². By what number were the dimensions multiplied?

_____

7. Sami put two trapezoids with the same dimensions together to make a parallelogram.

The formula for the area of a trapezoid is $A = \frac{1}{2}(b_1 + b_2)h$. Explain why the bases of a trapezoid need to be added in the formula.

8. GoDEEPER   A rectangular plastic bookmark has a triangle cut out of it. Use the diagram of the bookmark to complete the table.

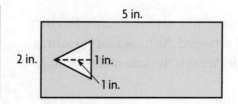

Area of Rectangle	Area of Triangle	Square Inches of Plastic in Bookmark

9. A trapezoid has an area of 32 in.^{2}. If the lengths of the bases are 6 in. and 6.8 in., what is the height?

_____ in.

10. A pillow is in the shape of a regular pentagon. The front of the pillow is made from 5 pieces of fabric that are congruent triangles. Each triangle has an area of 22 in.^{2}. What is the area of the front of the pillow?

_____ in.^{2}

11. Which expressions can be used to find the area of the trapezoid? Mark all that apply.

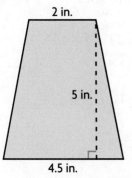

2 in.

5 in.

4.5 in.

(A) $\frac{1}{2} \times (5 + 2) \times 4.5$ (C) $\frac{1}{2} \times (5 + 4.5) \times 2$

(B) $\frac{1}{2} \times (2 + 4.5) \times 5$ (D) $\frac{1}{2} \times (6.5) \times 5$

12. Name the polygon and find its area. Show your work.

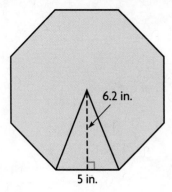

6.2 in.

5 in.

polygon: _____ area: _____

_____

13. A carpenter needs to replace some flooring in a house.

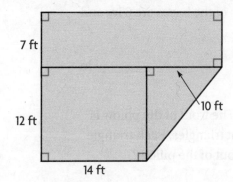

7 ft

12 ft

10 ft

14 ft

Select the expression that can be used to find the total area of the flooring to be replaced. Mark all that apply.

(A) 19×14

(C) $19 \times 24 - \frac{1}{2} \times 10 \times 12$

(B) $168 + 12 \times 14 + 60$

(D) $7 \times 24 + 12 \times 14 + \frac{1}{2} \times 10 \times 12$

14. Ava wants to draw a parallelogram on the coordinate plane. She plots these 3 points.

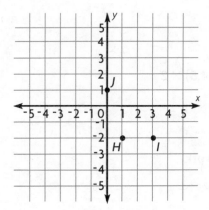

Part A

Find and label the coordinates of the fourth vertex, K, of the parallelogram. Draw the parallelogram.

Part B

What is the length of side JK? How do you know?

Name _____

15. Joan wants to reduce the area of her posters by one-third. Draw lines to match the original dimensions in the left column with the correct new area in the right column. Not all dimensions will have a match.

30 in. by 12 in. •		• 20 in.²
30 in. by 18 in. •		• 60 in.²
12 in. by 15 in. •		• 180 in.²
18 in. by 15 in. •		• 360 in.²

Personal Math Trainer

16. **THINK SMARTER ➕** Alex wants to enlarge a 4-ft by 6-ft vegetable garden by multiplying the dimensions of the garden by 2.

Part A

Find each area.

Area of original garden: _____

Area of enlarged garden: _____

Part B

Explain how the area of the original garden will be affected.

17. Suppose the point (3, 2) is changed to (3, 1) on this rectangle. What other point must change so the figure remains a rectangle? What is the area of the new rectangle?

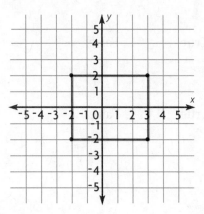

Point : _____ would change to _____.

The area of the new rectangle is _____ square units.

18. Look at the figure below. The area of the parallelogram and the areas of the two congruent triangles formed by a diagonal are related. If you know the area of the parallelogram, how can you find the area of one of the triangles?

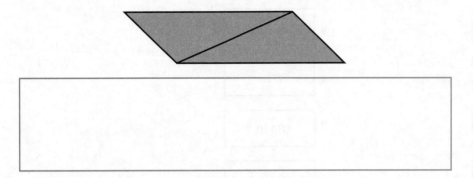

---

19. The roof of Kamden's house is shaped like a parallelogram. The base of the roof is 13 m and the area is 110.5 m². Choose a number and unit to make a true statement.

The height of the roof is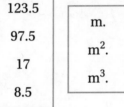

123.5
97.5
17
8.5

m.
m².
m³.

20. Eliana is drawing a figure on the coordinate grid. For numbers 20a–20d, select True or False for each statement.

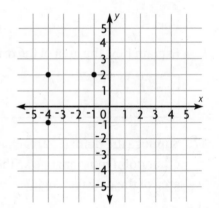

20a. The point (⁻1, 1) would be the fourth vertex of a square.　　○ True　　○ False

20b. The point (1, 1) would be the fourth vertex of a trapezoid.　　○ True　　○ False

20c. The point (2, ⁻1) would be the fourth vertex of a trapezoid.　　○ True　　○ False

20d. The point (⁻1, ⁻1) would be the fourth vertex of a square.　　○ True　　○ False